The Mathematics of Diffusion

CBMS-NSF REGIONAL CONFERENCE SERIES IN APPLIED MATHEMATICS

A series of lectures on topics of current research interest in applied mathematics under the direction of the Conference Board of the Mathematical Sciences, supported by the National Science Foundation and published by SIAM.

GARRETT BIRKHOFF, *The Numerical Solution of Elliptic Equations*
D. V. LINDLEY, *Bayesian Statistics, A Review*
R. S. VARGA, *Functional Analysis and Approximation Theory in Numerical Analysis*
R. R. BAHADUR, *Some Limit Theorems in Statistics*
PATRICK BILLINGSLEY, *Weak Convergence of Measures: Applications in Probability*
J. L. LIONS, *Some Aspects of the Optimal Control of Distributed Parameter Systems*
ROGER PENROSE, *Techniques of Differential Topology in Relativity*
HERMAN CHERNOFF, *Sequential Analysis and Optimal Design*
J. DURBIN, *Distribution Theory for Tests Based on the Sample Distribution Function*
SOL I. RUBINOW, *Mathematical Problems in the Biological Sciences*
P. D. LAX, *Hyperbolic Systems of Conservation Laws and the Mathematical Theory of Shock Waves*
I. J. SCHOENBERG, *Cardinal Spline Interpolation*
IVAN SINGER, *The Theory of Best Approximation and Functional Analysis*
WERNER C. RHEINBOLDT, *Methods of Solving Systems of Nonlinear Equations*
HANS F. WEINBERGER, *Variational Methods for Eigenvalue Approximation*
R. TYRRELL ROCKAFELLAR, *Conjugate Duality and Optimization*
SIR JAMES LIGHTHILL, *Mathematical Biofluiddynamics*
GERARD SALTON, *Theory of Indexing*
CATHLEEN S. MORAWETZ, *Notes on Time Decay and Scattering for Some Hyperbolic Problems*
F. HOPPENSTEADT, *Mathematical Theories of Populations: Demographics, Genetics and Epidemics*
RICHARD ASKEY, *Orthogonal Polynomials and Special Functions*
L. E. PAYNE, *Improperly Posed Problems in Partial Differential Equations*
S. ROSEN, *Lectures on the Measurement and Evaluation of the Performance of Computing Systems*
HERBERT B. KELLER, *Numerical Solution of Two Point Boundary Value Problems*
J. P. LASALLE, *The Stability of Dynamical Systems*
D. GOTTLIEB AND S. A. ORSZAG, *Numerical Analysis of Spectral Methods: Theory and Applications*
PETER J. HUBER, *Robust Statistical Procedures*
HERBERT SOLOMON, *Geometric Probability*
FRED S. ROBERTS, *Graph Theory and Its Applications to Problems of Society*
JURIS HARTMANIS, *Feasible Computations and Provable Complexity Properties*
ZOHAR MANNA, *Lectures on the Logic of Computer Programming*
ELLIS L. JOHNSON, *Integer Programming: Facets, Subadditivity, and Duality for Group and Semi-Group Problems*
SHMUEL WINOGRAD, *Arithmetic Complexity of Computations*
J. F. C. KINGMAN, *Mathematics of Genetic Diversity*
MORTON E. GURTIN, *Topics in Finite Elasticity*
THOMAS G. KURTZ, *Approximation of Population Processes*
JERROLD E. MARSDEN, *Lectures on Geometric Methods in Mathematical Physics*
BRADLEY EFRON, *The Jackknife, the Bootstrap, and Other Resampling Plans*
M. WOODROOFE, *Nonlinear Renewal Theory in Sequential Analysis*
D. H. SATTINGER, *Branching in the Presence of Symmetry*
R. TEMAM, *Navier–Stokes Equations and Nonlinear Functional Analysis*

MIKLÓS CSÖRGO, *Quantile Processes with Statistical Applications*
J. D. BUCKMASTER AND G. S. S. LUDFORD, *Lectures on Mathematical Combustion*
R. E. TARJAN, *Data Structures and Network Algorithms*
PAUL WALTMAN, *Competition Models in Population Biology*
S. R. S. VARADHAN, *Large Deviations and Applications*
KIYOSI ITÔ, *Foundations of Stochastic Differential Equations in Infinite Dimensional Spaces*
ALAN C. NEWELL, *Solitons in Mathematics and Physics*
PRANAB KUMAR SEN, *Theory and Applications of Sequential Nonparametrics*
LÁSZLÓ LOVÁSZ, *An Algorithmic Theory of Numbers, Graphs and Convexity*
E. W. CHENEY, *Multivariate Approximation Theory: Selected Topics*
JOEL SPENCER, *Ten Lectures on the Probabilistic Method*
PAUL C. FIFE, *Dynamics of Internal Layers and Diffusive Interfaces*
CHARLES K. CHUI, *Multivariate Splines*
HERBERT S. WILF, *Combinatorial Algorithms: An Update*
HENRY C. TUCKWELL, *Stochastic Processes in the Neurosciences*
FRANK H. CLARKE, *Methods of Dynamic and Nonsmooth Optimization*
ROBERT B. GARDNER, *The Method of Equivalence and Its Applications*
GRACE WAHBA, *Spline Models for Observational Data*
RICHARD S. VARGA, *Scientific Computation on Mathematical Problems and Conjectures*
INGRID DAUBECHIES, *Ten Lectures on Wavelets*
STEPHEN F. MCCORMICK, *Multilevel Projection Methods for Partial Differential Equations*
HARALD NIEDERREITER, *Random Number Generation and Quasi-Monte Carlo Methods*
JOEL SPENCER, *Ten Lectures on the Probabilistic Method, Second Edition*
CHARLES A. MICCHELLI, *Mathematical Aspects of Geometric Modeling*
ROGER TEMAM, *Navier–Stokes Equations and Nonlinear Functional Analysis, Second Edition*
GLENN SHAFER, *Probabilistic Expert Systems*
PETER J. HUBER, *Robust Statistical Procedures, Second Edition*
J. MICHAEL STEELE, *Probability Theory and Combinatorial Optimization*
WERNER C. RHEINBOLDT, *Methods for Solving Systems of Nonlinear Equations, Second Edition*
J. M. CUSHING, *An Introduction to Structured Population Dynamics*
TAI-PING LIU, *Hyperbolic and Viscous Conservation Laws*
MICHAEL RENARDY, *Mathematical Analysis of Viscoelastic Flows*
GÉRARD CORNUÉJOLS, *Combinatorial Optimization: Packing and Covering*
IRENA LASIECKA, *Mathematical Control Theory of Coupled PDEs*
J. K. SHAW, *Mathematical Principles of Optical Fiber Communications*
ZHANGXIN CHEN, *Reservoir Simulation: Mathematical Techniques in Oil Recovery*
ATHANASSIOS S. FOKAS, *A Unified Approach to Boundary Value Problems*
MARGARET CHENEY AND BRETT BORDEN, *Fundamentals of Radar Imaging*
FIORALBA CAKONI, DAVID COLTON, AND PETER MONK, *The Linear Sampling Method in Inverse Electromagnetic Scattering*
ADRIAN CONSTANTIN, *Nonlinear Water Waves with Applications to Wave-Current Interactions and Tsunamis*
WEI-MING NI, *The Mathematics of Diffusion*

WEI-MING NI
East China Normal University
Minhang, Shanghai
People's Republic of China
and
University of Minnesota
Minneapolis, Minnesota

The Mathematics of Diffusion

SOCIETY FOR INDUSTRIAL AND APPLIED MATHEMATICS
PHILADELPHIA

10 9 8 7 6 5 4 3 2 1

This research was supported in part by the National Science Foundation.

Library of Congress Cataloging-in-Publication Data

Ni, W.-M. (Wei-Ming)
The mathematics of diffusion / Wei-Ming Ni.
p. cm. -- (CBMS-NSF regional conference series in applied mathematics)
Includes bibliographical references and index.
ISBN 978-1-611971-96-5
1. Heat equation. I. Title.
QA377.N49 2011
515'.353--dc23

2011023014

獻給我的家鄉　我的祖國

To my homeland - Taiwan, China.

Contents

Preface

This book is an expanded version of the 10 CBMS lectures I delivered at Tulane University in May, 2010.

The main theme of this book is diffusion: from Turing's "diffusion-driven instability" in pattern formation to the interactions between diffusion and spatial heterogeneity in mathematical ecology. Along the way we will also discuss the effects of different boundary conditions—in particular, those of Dirichlet and Neumann boundary conditions.

On the dynamics aspect, in Chapters 1 and 2, we will start with the fundamental question of stabilization of solutions, including the rate of convergence. It seems interesting to note that the geometry of the underlying domain, although still far from being fully understood, plays a subtle role here.

It is well known that steady states play important roles in the dynamics of solutions to parabolic equations. In Chapter 3 we will focus on the qualitative properties of steady states, in particular, the "shape" of steady states and how it is related to the stability properties of steady states. This chapter is an updated version of relevant materials that appeared in an earlier survey article [N3].

In the second half of this book, we will first explore the interactions between diffusion and spatial heterogeneity, following the interesting theory developed mainly by Cantrell, Cosner, Lou, and others in mathematical ecology. Here it seems remarkable to note that even in the classical Lotka–Volterra competition-diffusion systems, the interaction of diffusion and spatial heterogeneity creates surprisingly different phenomena than its homogeneous counterpart. This is described in Chapter 4. Finally, in Chapter 5, we include several models beyond the usual diffusion, namely, various directed movements including chemotaxis and cross-diffusion models in population dynamics. This direction seems significant—from both modeling and mathematical points of view—one step into more sophisticated and realistic modeling with challenging and significant mathematical issues. In this regard, we would like to recommend the recent article [L3] to interested readers for a more thorough and complete survey.

Diffusion has been used extensively in many disciplines in science to model a wide variety of phenomena. Here we have included only a small number of models to illustrate the depth and breadth of the mathematics involved. The selection of the materials included here depends solely on my taste—not to reflect value judgement. One unfortunate omission is traveling waves, especially those with curved fronts. Interested readers are referred to the recent survey article of Taniguchi [Tn].

I wish to take this opportunity to thank the organizer of this CBMS conference, Xuefeng Wang, my old friend, Morris Kalka, and my colleagues and staffs at Tulane University for organizing this wonderful conference. I also wish to thank all the participants, some of whom came from faraway places, including China, Hong Kong, Japan, Korea, and Taiwan. It is a true pleasure to express my sincere appreciation to the five one-hour speakers, Chris Cosner, Manuel del Pino, Changfeng Gui, Kening Lu, and Juncheng Wei, for their inspiring lectures.

The manuscript was finalized when I was visiting Columbus, Ohio during the spring of 2011. I am grateful to Avner Friedman and Yuan Lou for providing such an ideal working environment for me to focus on mathematics. Finally, I wish to thank Adrian King-Yeung Lam for his help during the preparation of the final draft of this book—I have benefited much through our numerous discussions—and Yuan Lou for his invaluable insights regarding some of the materials presented here. This research was supported in part by the National Science Foundation.

Chapter 1

Introduction: The Heat Equation

1.1 Bounded Domains

We begin with pure diffusion, namely, the heat equation, on a bounded smooth domain

$$\begin{cases} u_t = \Delta u & \text{in } \Omega \times (0,\infty), \\ u(x,0) = u_0(x), & x \in \Omega, \end{cases}$$

where $u_t = \frac{\partial u}{\partial t}$, $\Delta = \sum_{i=1}^{n} \frac{\partial^2}{\partial x_i^2}$ is the usual Laplacian, and Ω is a bounded domain in $\mathbb{R}^n$ with smooth boundary $\partial\Omega$. Here $u(x,t)$ represents the heat distribution (the temperature) in Ω at time t.

Three types of boundary conditions are often considered: the Dirichlet boundary condition

$$u = 0 \quad \text{on } \partial\Omega,$$

the Neumann boundary condition

$$\partial_\nu u = 0 \quad \text{on } \partial\Omega,$$

where ν is the unit outward normal on $\partial\Omega$ and ∂_ν denotes the directional derivative $\frac{\partial}{\partial \nu}$, and the third boundary condition

$$\alpha u + \beta \partial_\nu u = 0 \quad \text{on } \partial\Omega,$$

where $\alpha, \beta > 0$ are two constants.

1.1.1 Dirichlet Boundary Condition

In the Dirichlet case

$$\begin{cases} u_t = \Delta u & \text{in } \Omega \times (0,\infty), \\ u = 0 & \text{on } \partial\Omega \times (0,\infty), \\ u(x,0) = u_0(x), & x \in \Omega, \end{cases} \tag{1.1}$$

it is well known that the solution stabilizes as t becomes large; that is,

$$u(x,t) \to 0 \tag{1.2}$$

as $t \to \infty$.

We are interested in the fundamental question of how fast the solution stabilizes in different domains; in other words, how does the convergence rate in (1.2) depend on the domain Ω? For instance, if $\Omega_1 \subseteq \Omega_2$, is the convergence rate in (1.2) faster for Ω_1?

To understand this problem, we start with the eigenvalues of the Laplace operator . Let $(0 <)\lambda_1(\Omega) < \lambda_2(\Omega) \leq \cdots$ be the eigenvalues of the Dirichlet eigenvalue problem

$$\begin{cases} \Delta\varphi + \lambda\varphi = 0 & \text{in } \Omega, \\ \varphi = 0 & \text{on } \partial\Omega, \end{cases} \tag{1.3}$$

and let $\varphi_1(\cdot\,;\Omega), \varphi_2(\cdot\,;\Omega), \ldots$ be the corresponding normalized eigenfunctions, i.e.,

$$\|\varphi_k(\cdot\,;\Omega)\|_{L^2(\Omega)} = 1.$$

(When there is no confusion, we will write λ_k for $\lambda_k(\Omega)$ and φ_k for $\varphi_k(\cdot\,;\Omega)$.) Then it is well known that the solution for (1.1) can be written as

$$u(x,t) = \sum_{k=1}^{\infty} \langle u_0, \varphi_k \rangle e^{-\lambda_k t} \varphi_k(x), \tag{1.4}$$

where $\langle u_0, \varphi_k \rangle$ denotes the L^2-inner product of u_0 and φ_k. Therefore, the convergence rate of $u \to 0$ is determined by the size of $\lambda_1(\Omega)$ in general. It turns out that all of the Dirichlet eigenvalues $\lambda_k(\Omega)$ satisfy the "Domain-Monotonicity Property":

Domain-Monotonicity Property: $\lambda_k(\Omega_1) \geq \lambda_k(\Omega_2)$ *for all* $k \geq 1$ *if* $\Omega_1 \subseteq \Omega_2$.

This implies that the solution u of the heat equation under the homogeneous Dirichlet boundary condition in (1.1) has the property that $u \to 0$ *faster if the domain* Ω *is smaller.*

1.1.2 Neumann Boundary Condition

For the Neumann problem

$$\begin{cases} u_t = \Delta u & \text{in } \Omega \times (0,\infty), \\ \partial_\nu u = 0 & \text{on } \partial\Omega \times (0,\infty), \\ u(x,0) = u_0(x), & x \in \Omega, \end{cases} \tag{1.5}$$

although the solution also stabilizes as t becomes large, i.e.,

$$u(x,t) \to \overline{u}_0 \equiv \frac{1}{|\Omega|} \int_\Omega u_0(x)\,dx \tag{1.6}$$

as $t \to \infty$, where $|\Omega|$ denotes the measure of Ω, the situation for the rate of convergence in (1.6) is drastically different from its Dirichlet counterpart. To describe the situation, similarly, we denote by $0 = \mu_1(\Omega) < \mu_2(\Omega) \leq \mu_3(\Omega) \leq \cdots$ the eigenvalues of

$$\begin{cases} \Delta\psi + \mu\psi = 0 & \text{in } \Omega, \\ \partial_\nu\psi = 0 & \text{on } \partial\Omega, \end{cases} \tag{1.7}$$

and we denote the corresponding normalized eigenfunction by $\psi_k(\cdot\,;\Omega)$. (Again, we suppress Ω in $\mu_k(\Omega)$ and $\psi_k(\cdot\,;\Omega)$ when there is no confusion.) Then the solution for (1.5) may be written as

$$u(x,t)=\sum_{k=1}^{\infty}\langle u_0,\psi_k\rangle e^{-\mu_k t}\psi_k(x)=\overline{u}_0+\left(\int_\Omega u_0\psi_2\right)e^{-\mu_2 t}\psi_2(x)+\cdots. \tag{1.8}$$

Thus the convergence rate of $u\to\overline{u}_0$ is decided by the size of $\mu_2(\Omega)$ in general. Differing from its Dirichlet counterpart, however, even the first positive eigenvalue $\mu_2(\Omega)$ does not enjoy the "Domain-Monotonicity Property."

There are many obvious examples to show that $\mu_2(\Omega_1)$ does not seem to have any relation with $\mu_2(\Omega_2)$ even if $\Omega_1\subseteq\Omega_2$. The following example seems, however, particularly illuminating. (We refer readers to [NW] for its detailed proof.)

Proposition 1.1 (see [NW, Theorem 2.1]). *Let*

$$\Omega_a=\begin{cases}[\,x\in\mathbb{R}^2\mid a<|x|<1\,] & \text{if } 0<a<1,\\ [\,x\in\mathbb{R}^2\mid 1<|x|<a\,] & \text{if } a>1.\end{cases}$$

Then $\mu_2(\Omega_a)$ is a strictly decreasing continuous function in $a\in(0,\infty)$, and

(i) $\lim_{a\to0^+}\mu_2(\Omega_a)=\mu_2(B_1)=\lim_{a\to\infty}a^2\mu_2(\Omega_a)$, *where B_1 is the unit disk in $\mathbb{R}^2$;*

(ii) $\lim_{a\to1}\mu_2(\Omega_a)=1.$

In an attempt to understand the phenomena presented in the proposition above, Ni and Wang [NW] propose the following notion of intrinsic distance and intrinsic diameter of Ω.

Definition 1.2.

(i) *Let P, Q be two points in Ω. Then the intrinsic distance between P and Q in Ω is defined as*

$$d_\Omega(P,Q)=\inf\{\,l(\mathcal{C})\mid \mathcal{C}\text{ is a continuous curve in }\Omega\text{ connecting }P\text{ and }Q\},$$

where $l(\mathcal{C})$ denotes the arc length of $\mathcal{C}$.

(ii) *The intrinsic diameter of Ω is defined as*

$$D(\Omega)=\sup\{d_\Omega(P,Q)\mid P,Q\in\Omega\}.$$

For example, the intrinsic diameter of the annulus Ω_a is the arc length of the minimal path in Ω_a between two antipodal points on the outer circle. (See Figure 1.1.)

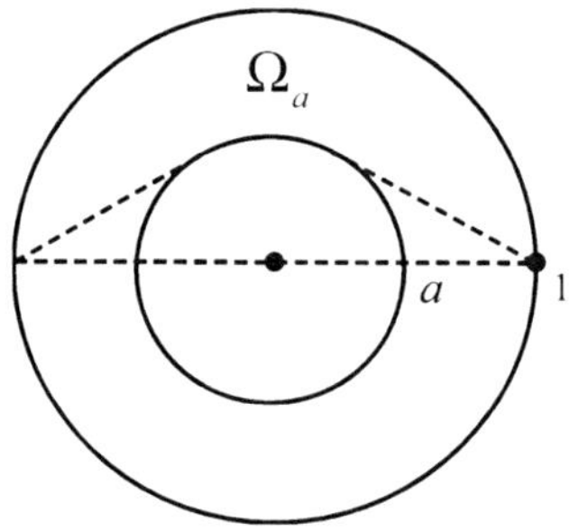

Figure 1.1.

To a certain extent, $D(\Omega)$ seems to measure the degree of difficulty for diffusion to *average* in Ω. Since $D(\Omega_a)$ is an increasing function of a, it seems reasonable to expect that $\mu_2(\Omega_a)$ decreases as a increases. However, for general domain Ω, the dependence of $\mu_2(\Omega)$ on $D(\Omega)$ is not as simple and clean—other factors, such as the geometry of the domain Ω, seem to come into play. For instance, the two domains Ω_1, Ω_2 in Figure 1.2 have the property that $\Omega_1 \subseteq \Omega_2$, but yet $\mu_2(\Omega_1)$ can be made arbitrarily small while $\mu_2(\Omega_2)$ can be made arbitrarily large.

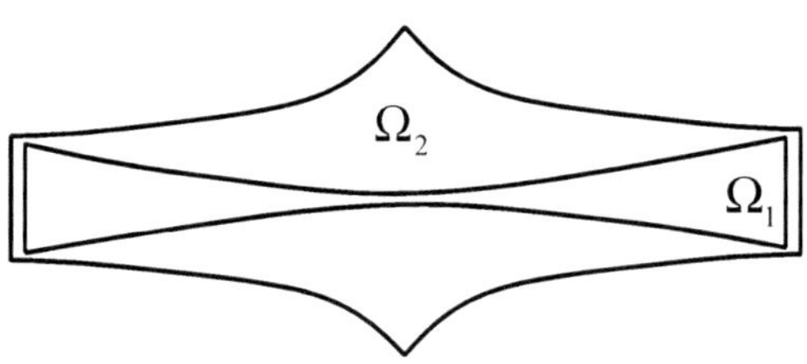

Figure 1.2.

To make this precise, we first establish the following preliminary result for "thin" domains under graphs of functions.

Proposition 1.3. *Let $x_2 = f(x_1)$ be a positive continuous function which is piecewise C^1 on the interval $[-1,1]$. Let Ω_f be the domain below the graph of f and above the interval $[-1,1]$. Then we have the following:*

(i) *$\mu_2(\Omega_{\varepsilon f})$ is a nonincreasing function of $\varepsilon > 0$.*

(ii) *$\lim_{\varepsilon\to 0} \mu_2(\Omega_{\varepsilon f}) = \mu_2^*$, where μ_2^* is the first positive eigenvalue of the Sturm–Liouville eigenvalue problem*

$$\begin{cases} (f(x)u')' + \mu f(x)u = 0, & x \in (-1,1), \\ u'(-1) = 0 = u'(1). \end{cases} \tag{1.9}$$

(iii) *Any eigenfunction $u_*(x)$ corresponding to μ_2^* is strictly monotone.*

(iv) *For small $\varepsilon > 0$, the eigenspace corresponding to $\mu_2(\Omega_{\varepsilon f})$ is one dimensional; and if f is even, then any eigenfunction $u^\varepsilon(x_1,x_2)$ corresponding to $\mu_2(\Omega_{\varepsilon f})$ is odd in x_1.*

(v) *Suppose in addition that f is piecewise C^2 on $[-1,1]$. Then any maximum point $x_\varepsilon^{\max}$ and minimum point $x_\varepsilon^{\min}$ of $u^\varepsilon(x_1,x_2)$ converge to the opposite lower corners of $\Omega_{\varepsilon f}$, i.e., $x_\varepsilon^{\max} \to (\pm 1,0)$ and $x_\varepsilon^{\min} \to (\mp 1,0)$ as $\varepsilon \to 0$.*

If $f \in C^3[-1,1]$, then part (ii) and the first half of part (iv) follow from [HR1, Theorem 4.3 and Proposition 4.9]. However, for our purposes we need to deal with more general f. We refer readers to [NW] for the detailed proof of Proposition 1.3.

Now we are ready to construct two families of domains; while all of them have essentially the same intrinsic diameter 2, the μ_2 of one family can be arbitrarily small, and that of the other can be arbitrarily large, as indicated in Figure 1.2.

Proposition 1.4.

(i) *If*

$$f_k(x) = e^{-k(1-|x|)}, \quad x \in [-1,1],$$

then for every fixed $c \in (0,1)$, we have

$$\mu_2(\Omega_{\varepsilon f_k}) = O(e^{-ck}) \quad \text{as } k \to \infty$$

for $\varepsilon > 0$ sufficiently small.

(ii) *Let*

$$f_k(x) = e^{-k|x|}.$$

Then, for any $k \geq 1$, there exists $\varepsilon_k > 0$ such that if $0 < \varepsilon < \varepsilon_k$,

$$\mu_2(\Omega_{\varepsilon f_k}) \geq \frac{k^2 + \pi^2}{4}.$$

Again, see [NW, Section 3] for the proof.

If we restrict our attention to convex domains in $\mathbb{R}^2$, more is known. First, note that *if Ω is convex, then the intrinsic diameter of Ω equals the diameter of Ω. Furthermore, it is known that*

$$\frac{\pi^2}{D^2(\Omega)} \leq \mu_2(\Omega) \leq \frac{4j_0^2}{D^2(\Omega)}, \tag{1.10}$$

where j_0 is the first zero of the Bessel function. It turns out that the estimates in (1.10) are sharp: *The lower bound can be approximated by a sequence of "thin" rectangles with the width approaching 0, while the upper estimate can be approximated by a sequence of "thin" rhombuses with the height approaching 0.*

The lower bound in (1.10) is due to [PW], and the upper bound is in [BB] by a probabilistic argument. Here we present a simple analytic proof by Wang [Wa2] inspired by the arguments in [BB].

Take $P, Q \in \partial\Omega$ with $dist(P,Q) = D(\Omega)$. Let l be the line perpendicular to $\overline{PQ}$ at its midpoint. Then l divides Ω into two parts, Ω_L and Ω_R. Let η_L be the first eigenvalue of the following problem with mixed boundary condition:

$$\begin{cases} \Delta v + \eta v = 0 & \text{in } \Omega_L, \\ v = 0 & \text{on } l \cap \Omega, \\ \partial_\nu v = 0 & \text{on } \partial\Omega_L \cap \partial\Omega, \end{cases} \tag{1.11}$$

i.e.,

$$\eta_L = \inf\left\{ \frac{\int_{\Omega_L} |\nabla v|^2}{\int_{\Omega_L} v^2} \;\middle|\; v \in H_1(\Omega_L), v = 0 \text{ on } l \cap \Omega \right\}.$$

Similarly, we define η_R.

First, we claim that

$$\mu_2(\Omega) \le \max\{\eta_L, \eta_R\}. \tag{1.12}$$

Let v_L and v_R be the positive normalized eigenfunctions corresponding to η_L and η_R, respectively. Extend v_L and v_R to the entire Ω by setting $v_L \equiv 0$ in $\Omega \setminus \Omega_L$ and $v_R \equiv 0$ in $\Omega \setminus \Omega_R$, respectively. Then $v_L, v_R \in H_1(\Omega)$. Choosing constants c_1, c_2 such that $\int_\Omega (c_1 v_L + c_2 v_R) = 0$, we have

$$\begin{aligned} \mu_2(\Omega) &\le \frac{\int_\Omega |\nabla(c_1 v_L + c_2 v_R)|^2}{\int_\Omega (c_1 v_L + c_2 v_R)^2} \\ &= \frac{c_1^2 \eta_L + c_2^2 \eta_R}{c_1^2 + c_2^2} \le \max\{\eta_L, \eta_R\} \end{aligned}$$

since v_L and v_R have disjoint supports, and our assertion (1.12) holds.

Next, observe that $4j_0^2/D^2(\Omega)$ is the first Dirichlet eigenvalue of the ball of radius $D(\Omega)/2$ centered at P, $B = B_{D(\Omega)/2}(P)$, and we denote by w the corresponding eigenfunction, positive and normalized with $\int_B w^2 = 1$. It is well known (e.g., by [GNN1]) that w is radially symmetric in B and decreasing in $r = |x - P|$. Setting

$$\widetilde{w}(x) = \begin{cases} w(x), & x \in \Omega \cap B, \\ 0, & x \in \Omega \setminus B, \end{cases}$$

we see that $\widetilde{w} \in H_1(\Omega_L)$ with $\widetilde{w} = 0$ on $l \cap \Omega$, and therefore

$$\eta_L \le \frac{\int_{\Omega_L} |\nabla \widetilde{w}|^2}{\int_{\Omega_L} \widetilde{w}^2} = \frac{\int_{B\cap\Omega_L} |\nabla w|^2}{\int_{B\cap\Omega_L} w^2}.$$

On the other hand, on $B \cap \Omega_L$, we have

$$\Delta w + 4\frac{j_0^2}{D^2(\Omega)} w = 0.$$

Thus

$$\begin{aligned} 4\frac{j_0^2}{D^2(\Omega)} \int_{B\cap\Omega_L} w^2 &= \int_{B\cap\Omega_L} |\nabla w|^2 - \int_{\partial(B\cap\Omega_L)} w \partial_\nu w \\ &\ge \int_{B\cap\Omega_L} |\nabla w|^2 \end{aligned}$$

as $\partial_\nu w \le 0$ on $\partial(B\cap\Omega_L)$, since the angle between ∇w and ν is bigger than or equal to 90°, by the convexity of Ω. (See Figure 1.3.) Hence $\eta_L \le 4j_0^2/D^2(\Omega)$. Similarly, $\eta_R \le 4j_0^2/D^2(\Omega)$ and the upper bound in (1.10) is now established.

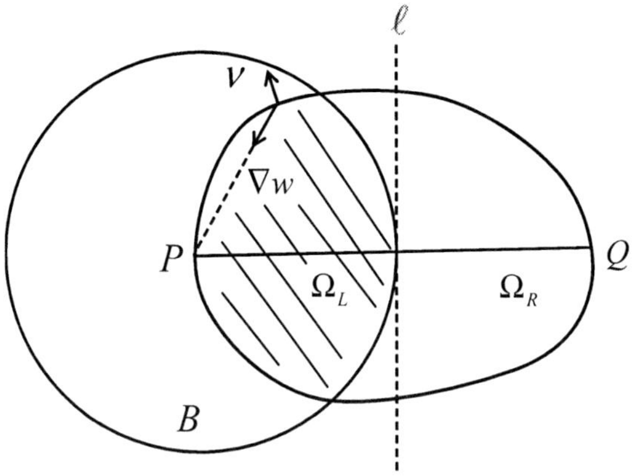

Figure 1.3.

1.1.3 Third Boundary Condition

For the third boundary condition

$$\begin{cases} u_t = \Delta u & \text{in } \Omega\times(0,\infty),\\ \alpha u+\beta\partial_\nu u = 0 & \text{on } \partial\Omega\times(0,\infty),\\ u(x,0)=u_0(x), & x\in\Omega, \end{cases}$$

the situation seems just as complicated as, if not more than, that of the Neumann case (1.5). Results obtained so far seem still preliminary—a direction worth pursuing.

1.1.4 Lipschitz Domains

Finally, we remark that although the domains in many of the examples mentioned in this chapter are *not* smooth domains, *all of them are Lipschitz domains. A Lipschitz domain Ω can always be approximated by smooth domains $\Omega_k, k = 1,2,\ldots,$ with the corresponding eigenvalues $\mu_2(\Omega_k)$ converging to $\mu_2(\Omega)$.* Some discussion in this direction already existed in [CH, pp. 421-423]; a more general result with rigorous proof is included in [NW, Appendix]. To make the above discussion precise, we follow [NW].

We will establish that any Lipschitz domain can be approximated by smooth domains in the sense described below so that μ_2 of the smooth domains converges to that of the Lipschitz domain.

Definition 1.5. *We say that a bounded domain $\Omega\subset\mathbb{R}^2$ is Lipschitz if for each point $P\in\partial\Omega$, there exist an open rectangular neighborhood R of P and a continuous and injective mapping $\eta: R\to\mathbb{R}^2$ such that*

(i) *both* η *and* η^{-1} *are Lipschitz continuous;*

(ii) $\eta(\Omega \cap R) \subset \{(y_1, y_2) \mid y_2 > 0\}$, *and* $\eta(\partial\Omega \cap R) \subset \{(y_1, y_2) \mid y_2 = 0\}$.

The pair (R, η) is called a *chart*. It follows from the above definition that there exist charts $\{(R_i, \eta_i)\}_{i=1}^N$ such that $\bigcup_{i=1}^N R_i \supset \partial\Omega$. The family $\{(R_i, \eta_i)\}_{i=1}^N$ is called an *atlas* of $\partial\Omega$. For a small positive δ, let $R_i^\delta = \{x \in R_i \mid dist(x, \partial R_i) > \delta\}$. Then $\{(R_i^\delta, \eta_i)\}_{i=1}^N$ is still an atlas of $\partial\Omega$ if δ is small enough—from now on, we fix such δ.

The other way to define the Lipschitz domain is to require $\partial\Omega$ to be locally the graph of a Lipschitz function. This is stronger than Definition 1.5. (See [Gd, pp. 7–9] for an example of a domain which is Lipschitz in the sense of the above definition, but its boundary is not the graph of any function near a boundary point.)

Definition 1.6. *A sequence of bounded domains* $\{\Omega_k\}$ *is said to converge to a bounded Lipschitz domain* Ω *if*

(i) *the characteristic functions* $\chi_{\Omega_k(x)} \to \chi_{\Omega(x)}$ *a.e. on* $\mathbb{R}^2$ *;*

(ii) *there exist an atlas* $\{(R_i, \eta_i)\}_{i=1}^N$ *of* $\partial\Omega$ *and mappings* $T_{ki} : R_i \to \mathbb{R}^2, k = 1, 2, \ldots, i = 1, \ldots, N$, *such that*

(a) *each* T_{ki} *is injective;*

(b) T_{ki} *and* T_{ki}^{-1} *are Lipschitz continuous with their Lipschitz norms bounded by* M, *which is independent of* k *and* i;

(c) $\bigcup_{i=1}^N T_{ki}(R_i^\delta) \supset \partial\Omega_k, T_{ki}(R_i \cap \partial\Omega) \subset \partial\Omega_k, T_{ki}(R_i \cap \overline{\Omega}^c) \subset \overline{\Omega}_k^c$.

Remark 1.7. *A condition, stronger than* (ii) *in* Definition 1.6 *but intuitively easier to check, is*

(ii′) *there exist open rectangles* $\{R_i\}_{i=1}^N$ *such that*

(a′) *the rectangles cover* $\partial\Omega$ *and* $\partial\Omega_k, k = 1, 2, \ldots$;

(b′) *for each* $i = 1, \ldots, N$, *there exists a Cartesian coordinate system* y_1, y_2 *such that the* y_1*-axis is parallel to one side of* R_i, $R_i \cap \partial\Omega$ *and* $R_i \cap \partial\Omega_k$ *are graphs of functions* $\psi_i(y_1)$ *and* $\psi_{ki}(y_1)$, *respectively, and* $R_i \cap \Omega = \{(y_1, y_2) \in R_i \mid y_2 > \psi_i(y_1)\}$ *and* $R_i \cap \Omega_k = \{(y_1, y_2) \in R_i \mid y_2 > \psi_{ki}(y_1)\}$;

(c′) $\psi_i(y_1)$ *and* $\psi_{ki}(y_1)$ *are Lipschitz continuous functions with their Lipschitz norms bounded by* M', *which is independent of* k *and* i.

Defining

$$T_{ki}(y_1, y_2) = (y_1, y_2 - \psi_i(y_1) + \psi_{ki}(y_1)), \quad T_{ki}^{-1}(z_1, z_2) = (z_1, z_2 + \psi_i(z_1) - \psi_{ki}(z_1)),$$

we easily see that (ii′) implies (ii).

The eigenvalue approximation theorem can now be stated as follows. (See [NW, Proposition 5.4] for a detailed rigorous proof.)

Proposition 1.8. *Suppose a sequence* $\{\Omega_k\}_{k=1}^\infty$ *of bounded Lipschitz domains converges to a bounded Lipschitz domain* Ω *in the sense of Definition* 1.6. *Then* $\lim_{k\to\infty} \mu_2(\Omega_k) = \mu_2(\Omega)$.

1.2 Entire Space $\mathbb{R}^n$

It is well known, by the example due to Tikhonov in 1935, that the Cauchy problem for the heat equation on $\mathbb{R}^n$ has more than one solution in general. To avoid such non-uniqueness situations, we shall restrict ourselves to the class of *bounded* functions on $\mathbb{R}^n$ in this section.

Here we again deal with the stabilization question: Consider the Cauchy problem

$$\begin{cases} u_t = u_{xx} & \text{in } \mathbb{R}\times(0,\infty), \\ u(x,0) = u_0(x), & x \in \mathbb{R}. \end{cases} \tag{1.13}$$

Suppose that u_0 is bounded in $\mathbb{R}$. Does the (bounded) solution $u(x,t)$ eventually stabilize?

Work in this direction goes back to as early as [Wi] in 1932. The following criterion can be found in, e.g., [E] in 1969.

Theorem 1.9. *The bounded solution $u(x,t)$ of* (1.13) *satisfies $\lim_{t\to\infty} u(0,t) = 0$ if and only if*

$$\lim_{R\to\infty} \frac{1}{2R}\int_{-R}^{R} u_0(y)\,dy = 0.$$

Moreover,

$$\lim_{t\to\infty}\left(\sup_{x\in\mathbb{R}} |u(x,t)|\right) = 0$$

if and only if

$$\limsup_{R\to\infty}\sup_{x\in\mathbb{R}} \frac{1}{2R}\left|\int_{x-R}^{x+R} u_0(y)\,dy\right| = 0.$$

An example, due to [CE], illustrates the situation further.

Example 1.10. *Let $u_0 \in C^\infty(\mathbb{R})$ be an even function with $|u_0|_{L^\infty} \le 1$ and*

$$u_0(x) = (-1)^k \text{ for } x \in [k! + 2^k, (k+1)! - 2^{k+1}].$$

Then the unique bounded solution $u(x,t)$ of (1.13) *satisfies*

$$\liminf_{t\to\infty} u(0,t) = -1 \quad \text{and} \quad \limsup_{t\to\infty} u(0,t) = 1.$$

For $n > 1$ and more general parabolic equations, much is also known. The following result is due to [K].

Theorem 1.11. *Let $u(x,t)$ be the bounded solution of the Cauchy problem*

$$\begin{cases} u_t = \sum_{i,j=1}^n \partial_{x_i}(a_{ij}(x,t)\partial_{x_j} u) & \text{in } \mathbb{R}^n \times (0,\infty), \\ u(x,0) = u_0(x), & x \in \mathbb{R}^n, \end{cases}$$

where u_0 is bounded and there exists a positive constant λ such that

$$\frac{1}{\lambda}|\xi|^2 \le \sum_{i,j=1}^n a_{ij}(x,t)\xi_i\xi_j \le \lambda|\xi|^2$$

for all $x, \xi \in \mathbb{R}^n, t \geq 0$. *Then*

$$\lim_{t\to\infty} u(x,t) = u_\infty(x)$$

exists if and only if

$$\lim_{R\to\infty} \frac{1}{|B_R(x)|} \int_{B_R(x)} u_0(y)\,dy = u_\infty(x),$$

where $B_R(x)$ *is the ball of radius* R *centered at* x; *moreover,* $u_\infty(x) \equiv \text{Constant}$.

We refer readers to [K] for the proof and a more detailed account of the history as well as other related results of this interesting problem.

Chapter 2

Dynamics of General Reaction-Diffusion Equations and Systems

In this chapter we continue to discuss various aspects of stabilization of solutions, but for *general semilinear reaction-diffusion equations or systems.* Different from the case of linear parabolic equations treated in Chapter 1, it is obvious that even with nice, e.g., bounded smooth, initial values, solutions of nonlinear equations could blow up at finite time. Therefore, the fundamental questions we are going to focus on in this chapter are the following:

(Q1) *Do bounded solutions necessarily converge?*

(Q2) *What necessary traits do steady states have to possess in order to be (locally) stable?*

From now on, we will deal mainly with homogeneous Neumann boundary conditions.

2.1 ω-Limit Set

We begin with single semilinear reaction-diffusion equations of the form

$$\begin{cases} u_t = d\Delta u + f(x,u) & \text{in } \Omega\times(0,T), \\ \partial_\nu u = 0 & \text{on } \partial\Omega\times(0,T), \\ u(x,0) = u_0(x), & x\in\Omega, \end{cases} \tag{2.1}$$

where the reaction term f is assumed to have sufficient smoothness, $T\in(0,\infty]$ gives the maximal internal of existence in time of the solution $u(x,t)$, and the constant $d>0$ denotes the rate of diffusion.

First, it is well known that the natural "energy" functional associated with (2.1),

$$J[u](t) = \int_\Omega \left[\frac{d}{2}|\nabla u(x,t)|^2 - F(x,u(x,t))\right] dx, \tag{2.2}$$

where $F(x,u) = \int_0^u f(x,s)ds$, is always strictly decreasing along a trajectory.

Proposition 2.1. $\frac{d}{dt}J[u](t) < 0$ *for all* $t \in (0,T)$, *provided that* u_0 *is not a steady state of* (2.1).

Proof. It is straightforward to verify that

$$\frac{d}{dt}J[u](t) = -\int_\Omega u_t^2(x,t)\,dx \le 0.$$

Thus it only remains to show that *if for some* $t_0 \ge 0$, $u_t(x,t_0) \equiv 0$ *for all* $x \in \Omega$, *then* $u_t(x,t) \equiv 0$ *for all* $x \in \Omega$ *and for all* $t \ge 0$. To achieve this, we consider the following two cases separately.

(i) $t > t_0$: Let $w(x,t) = u(x,t) - u(x,t_0)$. Then

$$\begin{cases} w_t = d\Delta w + c(x,t)w & \text{in } \Omega \times (t_0,T), \\ \partial_\nu w = 0 & \text{on } \partial\Omega \times (t_0,T), \\ w(x,t_0) = 0, & x \in \Omega, \end{cases}$$

where

$$c(x,t) = \begin{cases} \dfrac{f(x,u(x,t)) - f(x,u(x,t_0))}{u(x_1,t) - u(x,t_0)} & \text{if } u(x,t) \ne u(x,t_0), \\ \partial_u f(x,u(x,t_0)) & \text{if } u(x,t) = u(x,t_0), \end{cases}$$

is a smooth function. Now, the standard Maximum Principle implies that $w(x,t) \equiv 0$ for all $x \in \Omega$ and $t \ge t_0$.

(ii) $0 \le t \le t_0$: This is the "backward uniqueness" for parabolic equations. We proceed as follows. Setting $w(x,t) = u(x,t) - u(x,t_0)$ for $x \in \Omega$ and $0 < t < t_0$, we have that

$$\begin{cases} w_t = d\Delta w + c(x,t)w & \text{in } \Omega \times (0,t_0), \\ \partial_\nu w = 0 & \text{on } \partial\Omega \times (0,t_0), \\ w(x,t_0) = 0, & x \in \Omega, \end{cases} \tag{2.3}$$

where again

$$c(x,t) = \begin{cases} \dfrac{f(x,u(x,t)) - f(x,u(x,t_0))}{u(x,t) - u(x,t_0)} & \text{if } u(x,t) \ne u(x,t_0), \\ \partial_u f(x,u(x,t)) & \text{if } u(x,t) = u(x,t_0), \end{cases} \tag{2.4}$$

is a smooth function on $\Omega \times (0,t_0]$.

Suppose for some $t_1 \in (0,t_0)$, $w(x,t_1) \not\equiv 0$ in Ω. Then the function

$$\Lambda(t) = \frac{\int_\Omega |\nabla w|^2\,dx}{\int_\Omega w^2\,dx} \tag{2.5}$$

is well defined in a neighborhood of t_1. Set for simplicity $d = 1$. Straightforward computation shows that

$$\begin{aligned}\Lambda'(t) &= -\left(\frac{2}{\int_\Omega w^2}\right)\int_\Omega (\Delta w + cw)(\Delta w + \Lambda w)\,dx \\ &= -\left(\frac{2}{\int_\Omega w^2}\right)\int_\Omega [\,(\Delta w + \Lambda w)^2 + cw(\Delta w + \Lambda w)\,] \\ &\le \frac{2}{\int_\Omega w^2}\int_\Omega \frac{1}{4}c^2 w^2 \le \frac{1}{2}C\end{aligned}$$

since

$$\int w(\Delta w + \Lambda w) = 0,$$

where $C = \sup_{0\le t\le t_0} \|c(\cdot,t)\|^2_{L^\infty(\Omega)}$. Thus $\Lambda(t)$ must be bounded in $[t_1, \widetilde{t_0})$, where $\widetilde{t_0} \le t_0$ is the first instance t in $(t_1, t_0]$ at which $w(\cdot,t) \equiv 0$ in Ω. On the other hand,

$$\begin{aligned}\frac{d}{dt}\left(\ln \frac{1}{\int_\Omega w^2\,dx}\right) &= 2\,\Lambda(t) - 2\frac{\int_\Omega cw^2}{\int_\Omega w^2} \\ &\le 2\Lambda(t) + 2\|c(\cdot,t)\|_{L^\infty(\Omega)}.\end{aligned}$$

Hence, the function $\ln(\frac{1}{\int_\Omega w^2\,dx})$ is bounded as long as $\Lambda(t)$ remains bounded, which prevents $\int_\Omega w^2\,dx$ from vanishing in $(t_1, \widetilde{t_0}]$, contradicting the assumption that $w(\cdot,\widetilde{t_0}) \equiv 0$ in Ω. Therefore w must vanish in $\Omega \times [0, t_0]$, and our proof is complete. □

Proposition 2.1 plays an important role in understanding the limiting behavior of solutions of (2.1).

Let $u(x,t;u_0)$ be the solution of (2.1) with the initial value $u_0(x)$. We define its *ω-limit* set as follows:

$$\omega[u_0] = \bigcap_{\tau\ge 0} \overline{\{u(\cdot,t;u_0)\,|\; t \ge \tau\}},$$

where the closure can be taken either in $L^\infty(\Omega)$ or in $C^2(\overline{\Omega})$ due to the parabolic regularity estimates. It is clear that if $v \in \omega[u_0]$ is an isolated element, then $\omega[u_0] = \{v\}$.

Proposition 2.2. *Let $v \in \omega[u_0]$. Then v is a steady state for* (2.1).

Proof. Let $v \in \omega[u_0]$. Then there is a sequence $t_k \to \infty$ such that $u(\cdot,t_k;u_0) \to v$. By the continuous dependence on initial values we have, for any $\widetilde{t} > 0$,

$$u(\cdot, t_k + \widetilde{t}; u_0) \to u(\cdot, \widetilde{t}; v).$$

Hence

$$J[u(\cdot, t_k + \widetilde{t}; u_0)] \to J[u(\cdot, \widetilde{t}; v)].$$

Since $J[u(\cdot,t;u_0)]$ is monotonically decreasing in t by Proposition 2.1, it follows that

$$\begin{aligned}J[u(\cdot,\widetilde{t};v)] &= \lim_{k\to\infty} J[u(\cdot, t_k + \widetilde{t}; u_0)] \\ &= \lim_{k\to\infty} J[u(\cdot, t_k; u_0)] = J[v]\end{aligned}$$

for any $\widetilde{t} > 0$. By Proposition 2.1 again, we conclude that v is a steady state. □

2.2 Dynamics of Single Equations

2.2.1 Stabilization

We now take up questions (Q1) and (Q2) proposed at the beginning of this chapter for solutions of (2.1).

Note that Proposition 2.1 already excludes the possibility of periodic solutions. However, it is still remarkable that, in as early as 1968, Zelenyak [Z] studied question (Q1) and showed that *in case* $n = 1$, i.e., when Ω is a bounded interval in $\mathbb{R}$, *bounded solutions of* (2.1) *always converge*. (The same result was reproduced 10 years later in 1978 by Matano [Ma1].) For $n \geq 2$, question (Q1) was again answered affirmatively by Simon in 1983 [Si] *if* f *is analytic in* u. The C^∞-case, however, turns out to be false. There is a considerable amount of literature in constructing such examples—bounded nonconvergent solutions of (2.1)—and in understanding such phenomena under various titles as: realizations of vector fields on invariant manifolds, semilinear heat equations with bounded nonconvergent solutions, trajectories with multidimensional limit sets, existence of chaotic dynamics, etc. We refer interested readers to the survey [Po]. Here we shall conclude this subsection with two remarks.

(i) *If one allows the reaction term in* (2.1) *to depend on* ∇u *as well, then neither* Proposition 2.1 *nor* Proposition 2.2 *holds, and the dynamics become very complicated. More precisely, consider*

$$\begin{cases} u_t = d\Delta u + f(x,u,\nabla u) & \textit{in } \Omega \times (0,T), \\ \partial_\nu u = 0 & \textit{on } \partial\Omega \times (0,T), \\ u(x,0) = u_0(x) & \textit{in } \Omega. \end{cases} \tag{2.6}$$

Then the dynamics of any ODE system can be embedded into the dynamics of (2.6) *for some suitably chosen* $f(x,u,\nabla u)$*; in particular, periodic solutions or even chaotic behavior can be present.* (See [Po].)

(ii) *On the other hand, if the reaction term in* (2.1) *is independent of* x*, then it is not known whether there is a bounded nonconvergent solution. More explicitly, consider*

$$\begin{cases} u_t = d\Delta u + f(u) & \textit{in } \Omega \times (0,T), \\ \partial_\nu u = 0 & \textit{on } \partial\Omega \times (0,T), \\ u(x,0) = u_0(x) & \textit{in } \Omega. \end{cases} \tag{2.7}$$

It is an *open* problem *whether* (2.7) *has a bounded nonconvergent solution for some smooth* f.

2.2.2 Stability and Linearized Stability

We now turn to question (Q2) to study the stability and instability properties of steady states.

Let v be a steady state of (2.1); i.e.,

$$\begin{cases} d\,\Delta v + f(x,v) = 0 & \text{in } \Omega, \\ \partial_\nu v = 0 & \text{on } \partial\Omega. \end{cases} \tag{2.8}$$

Definition 2.3. *v is stable if for any $\varepsilon > 0$, there exists $\delta > 0$ such that $\|u_0 - v\|_{L^\infty(\Omega)} < \delta$ implies*

$$\|u(\cdot,t;u_0) - v\|_{L^\infty(\Omega)} < \varepsilon$$

for all $t > 0$, where $u(\cdot,t;u_0)$ is the solution of (2.1).

Definition 2.4. *v is asymptotically stable if there exists $\delta > 0$ such that $\|u_0 - v\|_{L^\infty(\Omega)} < \delta$ implies*

$$\lim_{t\to\infty} \|u(\cdot,t;u_0) - v\|_{L^\infty(\Omega)} = 0.$$

We first make a connection between the stability defined above and the *linearized stability at v*. The first eigenvalue μ_1 of the linearized problem of (2.8) is given by

$$\mu_1 = \inf\left\{ \frac{\int_\Omega \left[d\,\|\nabla\psi\|^2 - f_u(x,v)\,\psi^2 \right]}{\int_\Omega \psi^2} \,\middle|\, \psi \in H_1(\Omega)\setminus\{0\} \right\} \tag{2.9}$$

and the corresponding normalized eigenfunction ψ_1 satisfies

$$\begin{cases} d\Delta\psi_1 + f_u(x,v)\psi_1 + \mu_1\psi_1 = 0 & \text{in } \Omega, \\ \partial_\nu \psi_1 = 0 & \text{on } \partial\Omega. \end{cases} \tag{2.10}$$

It is well known that μ_1 is simple and $\psi_1 > 0$ in $\overline{\Omega}$.

Lemma 2.5. *If $\mu_1 < 0$, then v is unstable.*

Proof. Suppose, for contradiction, that v is stable. By the continuity of $f_u(x,\cdot)$, there exists an $\varepsilon > 0$ small such that

$$|f_u(x,v(x)+h) - f_u(x,v(x))| < \frac{|\mu_1|}{2} \tag{2.11}$$

for all $x \in \overline{\Omega}$ and for all $|h| < \varepsilon$. From Definition 2.3 we can choose an initial value u_0 such that

$$v(x) < u_0(x) < v(x) + \delta$$

for all $x \in \overline{\Omega}$. Then for all $t > 0$

$$\|u(\cdot,t;u_0) - v\|_{L^\infty(\Omega)} < \varepsilon. \tag{2.12}$$

Observe that it always holds that $u(\cdot,t;u_0) > v$ in $\overline{\Omega}$ by the Maximum Principle.

Now we set

$$g(t) = \int_\Omega [u(x,t;u_0) - v(x)]\psi_1(x)\,dx$$

and differentiate

$$\begin{aligned}
g'(t) &= \int_\Omega u_t \psi_1 \, dx \\
&= \int_\Omega [\, d\Delta u + f(x,u)]\psi_1(x)\, dx \\
&= \int_\Omega \big[\, d\Delta(u-v) + (f(x,u) - f(x,v))\big]\psi_1 \, dx \\
&= \int_\Omega (u-v)\left[\, d\Delta\psi_1 + \frac{f(x,u)-f(x,v)}{u-v}\psi_1\right] dx \\
&= \int_\Omega (u-v)\left[\frac{f(x,u)-f(x,v)}{u-v} - f_u(x,v) - \mu_1\right]\psi_1 \, dx \\
&\geq \frac{|\mu_1|}{2} g(t)
\end{aligned}$$

by (2.11) and (2.12). Therefore

$$g(t) \geq g(0)\exp\left(\frac{|\mu_1|}{2}t\right) \to \infty$$

as $t \to \infty$, a contradiction. □

2.2.3 Stability for Autonomous Equations

The main theorem in this subsection basically says that *single "autonomous" reaction-diffusion equations do not support interesting patterns*, at least for convex domains. More precisely, consider the special case of (2.1), where f does not depend on x explicitly,

$$\begin{cases} u_t = d\Delta u + f(u) & \text{in } \Omega \times (0,T), \\ \partial_\nu u = 0 & \text{on } \partial\Omega \times (0,T), \\ u(x,0) = u_0(x) & \text{in } \Omega, \end{cases} \tag{2.13}$$

and let v be a steady state; i.e.,

$$\begin{cases} d\Delta v + f(v) = 0 & \text{in } \Omega, \\ \partial_\nu v = 0 & \text{on } \partial\Omega. \end{cases} \tag{2.14}$$

Then we have the following.

Theorem 2.6. *If Ω is convex and v is stable, then v must be a constant.*

In other words, for autonomous equations on convex domains, *"stability implies triviality!"*

This result is due to Casten and Holland [CaH] in 1977. (Again, Matano [Ma2] came up with the same theorem later.) The convexity of Ω is crucial and is needed in the following lemma.

Lemma 2.7. *Let Ω be convex, and let $w \in C^2(\overline{\Omega})$ with $\partial_\nu w = 0$ on $\partial\Omega$. Then $\partial_\nu |\nabla w|^2 \leq 0$ on $\partial\Omega$.*

Proof. Let $P \in \partial\Omega$ be an arbitrary but fixed point. After rotation and translation, we may assume that P is located at the origin and the boundary $\partial\Omega$ near P may be represented by $x_n = \varphi(x')$, where $x' = (x_1, \ldots, x_{n-1})$ with

$$\nabla' \varphi(0) = \langle \partial_{x_1} \varphi(0), \ldots, \partial_{x_{n-1}} \varphi(0) \rangle = 0$$

and the outer normal $\nu = \langle 0, \ldots, 0, -1 \rangle$. By the convexity of Ω at 0, the Hessian $(\partial^2_{x_i x_j} \varphi(0))$ is *nonnegative definite.*

Since

$$\nu = \frac{\langle \nabla' \varphi, -1 \rangle}{|\langle \nabla' \varphi, -1 \rangle|}$$

on $\partial\Omega$ near 0, we have

$$0 = \frac{\partial w}{\partial \nu} = \nabla w \cdot \nu = \frac{\sum_{i=1}^{n-1} (\partial_{x_i} w \cdot \partial_{x_i} \varphi - \partial_{x_n} w)}{\sqrt{\sum_{j=1}^{n-1} (\partial_{x_j} \varphi)^2 + 1}}.$$

Hence

$$\partial_{x_n} w = \sum_{i=1}^{n-1} \partial_{x_i} w \cdot \partial_{x_i} \varphi \tag{2.15}$$

on $\partial\Omega$ near 0, i.e., on $(x', \varphi(x'))$ for x' small. Differentiating (2.15) with respect to x_j, $1 \leq j \leq n-1$, we have

$$\partial^2_{x_n x_j} w + \partial^2_{x_n x_n} w \cdot \partial_{x_j} \varphi = \sum_{i=1}^{n-1} (\partial^2_{x_i x_j} w \cdot \partial_{x_i} \varphi + \partial^2_{x_i x_n} w \cdot \partial_{x_j} \varphi \cdot \partial_{x_i} \varphi + \partial_{x_i} w \cdot \partial^2_{x_i x_j} \varphi)$$

on $(x', \varphi(x'))$ near $x' = 0$. Thus, for $1 \leq j \leq n-1$,

$$\partial^2_{x_n x_j} w(0) = \sum_{i=1}^{n-1} \partial_{x_i} w(0) \cdot \partial^2_{x_i x_j} \varphi(0) \tag{2.16}$$

since $\nabla' \varphi(0) = 0$. Now

$$\begin{aligned} \partial_\nu |\nabla w|^2(0) &= -\partial_{x_n} |\nabla w|^2(0) \\ &= -2 \left[\left(\sum_{j=1}^{n-1} \partial_{x_j} w \cdot \partial^2_{x_j x_n} w \right) + \partial_{x_n} w \cdot \partial_{x_n x_n} w \right] (0) \\ &= -2 \sum_{i,j=1}^{n-1} \partial_{x_i} w(0) \cdot \partial_{x_j} w(0) \cdot \partial^2_{x_i x_j} \varphi(0) \leq 0 \end{aligned}$$

by (2.16) and the fact that

$$\partial_{x_n} w(0) = -\partial_\nu w(0) = 0.$$

This completes the proof. □

We are now ready for the following proof.

Proof of Theorem 2.6. Suppose, for contradiction, that v is not a constant. Differentiating the equation in (2.14), we have

$$\Delta(\partial_{x_k} v) + f'(v)(\partial_{x_k} v) = 0$$

in Ω. Multiplying the equation by $\partial_{x_k} v$ and integrating over Ω, we obtain

$$0 = \int_\Omega \left[-|\nabla(\partial_{x_k} v)|^2 + f'(v)(\partial_{x_k} v)^2 \right] + \frac{1}{2} \int_{\partial\Omega} \partial_\nu [(\partial_{x_k} v)^2].$$

Summing up over $k = 1, \ldots, n$, we have

$$0 = \sum_{k=1}^n \int \left[-|\nabla(\partial_{x_k} v)|^2 + f'(v)(\partial_{x_k} v)^2 \right] + \frac{1}{2} \int_{\partial\Omega} \partial_\nu [(\partial_{x_k} v)^2].$$

Thus

$$0 \geq \sum_{k=1}^n \int_\Omega \left[|\nabla(\partial_{x_k} v)|^2 - f'(v)(\partial_{x_k} v)^2 \right]$$

by Lemma 2.7. Since $\nabla v \not\equiv 0$, $\mu_1 \leq 0$. On the other hand, $\mu_1 \geq 0$ by the stability of v and Lemma 2.5. Therefore, $\mu_1 = 0$ and $\partial_{x_k} v = c_k \psi_1$, as μ_1 is a simple eigenvalue, where ψ_1 is the corresponding first eigenfunction; i.e., $\nabla v = \overrightarrow{c}\,\psi_1$, where $\overrightarrow{c} = \langle c_1, \ldots, c_n \rangle$. After a rotation, we may assume $\overrightarrow{c} = \langle c, 0, \ldots, 0 \rangle$, where $c > 0$; i.e., $v(x) = v(x_1)$, a single variable function. Now, (2.14) reduces to

$$\begin{cases} dv'' + f(v) = 0 & \text{in } (a,b), \\ v'(a) = v'(b) = 0, \end{cases} \tag{2.17}$$

where Ω lies between the hyperplanes $x_1 = a$ and $x_1 = b$. Since $\partial_{x_1} v = c\psi_1$ and $\partial_\nu \psi_1 = 0$ on $\partial\Omega$, we have $v''(a) = 0$, and thus $f(v(a)) = 0$ by (2.17). It then follows from the uniqueness of solutions to initial value problems for ODEs that $v \equiv v(a)$ is the solution, contradicting our assumption that v is not a constant. Our proof is now complete. □

Remark 2.8. *In* [Ma2], *a stable nonconstant steady state for* (2.14) *was constructed for a dumbbell-shaped domain* Ω, *which shows that the convexity of* Ω *is essential for Theorem* 2.6 *to hold.*

Remark 2.9. *For Dirichlet problems, the notion of stability and linearized eigenvalue can be defined in a similar fashion and the counterpart of Lemma* 2.5 *can also be obtained in a similar way. However, the general situation is quite different.*

For example, it is easy to show that, via upper- and lower-solution methods, the Dirichlet problem

$$\begin{cases} \Delta v + v^p = 0 & \textit{in } \Omega, \\ v = 0 & \textit{on } \partial\Omega \end{cases} \tag{2.18}$$

has a unique positive stable solution if $0 < p < 1$ *and has no positive stable solution if* $p > 1$. *Furthermore, it is not hard to show that for general reaction term* $f(v)$, *stable solutions of*

$$\begin{cases} \Delta v + f(v) = 0 & \text{in } \Omega, \\ v = 0 & \text{on } \partial\Omega \end{cases} \tag{2.19}$$

must not change sign if Ω *is a ball or an annulus. However, examples of stable solutions for* (2.19) *which change sign in* Ω *have been constructed even for convex domains* [Sw], [DG].

Finally, we remark that when $n = 1$, *under general Dirichlet boundary conditions, the stability question has been studied in* [Mg].

2.2.4 Stability for Nonautonomous Equations

In spatially *inhomogeneous* cases, i.e., when f depends on x (as in (2.8)), then Theorem 2.6 no longer holds. In the work of Nakashima [Na1, Na2], among other things, *stable steady states with transition layers near nondegenerate minimum points* of $h(x)$ are constructed for the inhomogeneous Allen–Cahn equation in one dimension,

$$\begin{cases} \varepsilon u_t = \varepsilon u_{xx} + \frac{1}{\varepsilon} h^2(x) u(1-u^2) & \text{in } (0,1)\times(0,\infty), \\ u_x(0,t) = u_x(1,t) = 0 & \text{in } (0,\infty), \end{cases} \tag{2.20}$$

where $h > 0$, provided that $\varepsilon > 0$ is sufficiently small. In fact, *very precise and complete* results concerning steady states with layers near maximum and/or minimum points of h are obtained in [Na1, Na2]. More precisely, Nakashima proved in [Na1, Na2] that *if* $h > 0$ *on* $[0,1]$ *and nondegenerate at its extremum points, then, for* $\varepsilon > 0$ *sufficiently small, there exists a steady state of* (2.20) *with any (given) number of clustering layers at an arbitrarily chosen subset of local maximum points of h and a single layer at an arbitrarily chosen subset of local minimum points of h. Moreover, layer solutions near minimum points of h are stable, while layer solutions near maximum points of h are unstable.* (See Figure 2.1.) The multidimensional counterpart of these results are also treated by Li and Nakashima [LNa].

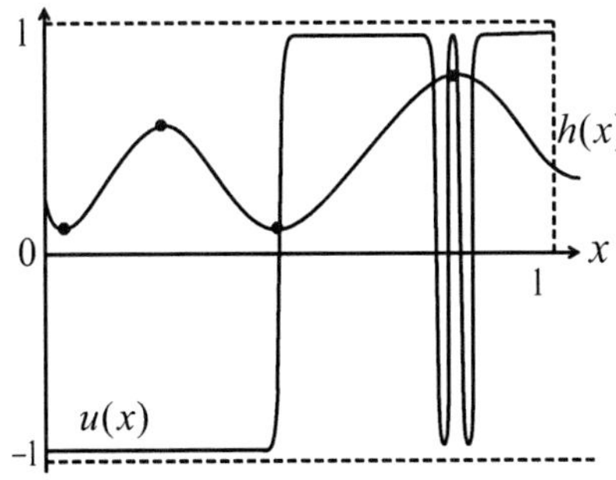

Figure 2.1.

Generalizing Theorem 2.6 to the nonautonomous case in the multidimensional domain remains largely open, although progress has been made recently. In [LNN], the following equation on a ball or an annulus has been considered:

$$\begin{cases} u_t = d\Delta u + h^2(x) f(u) & \text{in } \Omega \times (0,T), \\ \partial_\nu u = 0 & \text{on } \partial\Omega \times (0,T), \end{cases} \tag{2.21}$$

where h is a *positive smooth* function on $\overline{\Omega}$, and $T \leq \infty$.

Theorem 2.10. *Suppose that $n \geq 2$, $\Omega = B_R\,(B_{R_1} \backslash \overline{B_{R_2}}$, resp.), and $h(x) = h(r)$, where $r = |x|$. Let v be a nonconstant steady state of* (2.21).

(i) *If v is not radially symmetric, then v is unstable.*

(ii) *If v is radially symmetric and h satisfies*

$$\left[r^{n-1} \left(\frac{1}{r^{n-1} h} \right)' \right]' \geq 0 \tag{2.22}$$

in $B_R\,(B_{R_1} \backslash B_{R_2}$, resp.), then v is unstable.

Of course the autonomous case $h \equiv 1$ is included in (2.22). Furthermore, a counterexample of a *stable nonconstant radially symmetric steady state of* (2.21) was constructed in [LNN, Proposition 2.1] by using the earlier work of Nakashima [Na2] on (2.20) when the condition (2.22) is *not* satisfied.

The strategy in proving Theorem 2.10 is again to produce a negative eigenvalue μ_1 for the linearized eigenvalue problem at v, as in Subsection 2.2.2. First, if v is *not* radial, then

$$v_\theta = (x_1 \partial_{x_2} - x_2 \partial_{x_1}) v \not\equiv 0$$

is an eigenfunction corresponding to the 0 eigenvalue in the linearized problem at v. Since v_θ changes sign in Ω, it cannot be the first eigenfunction, and thus 0 is *not* the first eigenvalue.

If v is radially symmetric and nonconstant, the conclusion follows from the following lemma, which is interesting in its own right and will be useful when we deal with systems later.

Lemma 2.11. *Let v be a nonconstant radially symmetric steady state of* (2.21) *with $\Omega = B_R$ or $B_{R_1} \backslash \overline{B_{R_2}}$, and let $h(x) = h(r)$, $r = |x|$.*

(i) *If* (2.22) *holds in Ω, then there is at least one negative eigenvalue for the linearized problem at v.*

(ii) *If h satisfies*

$$\left[r^{n-1} \left(\frac{1}{r^{n-1} h} \right)' \right]' \geq \frac{n-1}{r^2 h} \tag{2.23}$$

in Ω, then there is at least one negative eigenvalue with multiplicity bigger than 1 for the linearized problem at v.

The proof involves tedious calculations, and we refer interested readers to [LNN, Lemma 2.4] for details.

2.3 Dynamics of 2×2 Systems and Their Shadow Systems

While we have fairly good knowledge of the dynamics of single equations (described in the previous section of this chapter), our understanding of the dynamics of 2×2 systems remains limited to this date. The counterparts of Theorems 2.6 and 2.10 are not known, except for some very special cases where the nonlinearities in the 2×2 systems have specific structures, e.g., gradient structure (such as the Ginzburg–Landau system), or skew-gradient structure (as some very special cases of the Gierer–Meinhardt system). (See, e.g., [JM2], [Ya], [KW].) Furthermore, most of the known examples have similar conclusions—that is, "stability implies triviality"—as in the single equations. On the other hand, we already know that 2×2 systems do allow many striking patterns that are stable. Therefore, instead of describing those specific examples which allow no *stable nontrivial steady states*, we will present general theorems by varying the diffusion rates in 2×2 systems.

In this generality, most of our understanding is limited to autonomous 2×2 systems. Thus, we shall restrict ourselves mostly to the following autonomous systems in this section and make remarks on more general (e.g., nonautonomous) systems whenever appropriate:

$$\begin{cases} U_t = d_1 \Delta U + f(U,V) & \text{in } \Omega \times (0,T), \\ V_t = d_2 \Delta V + g(U,V) & \text{in } \Omega \times (0,T), \\ \partial_\nu U = \partial_\nu V = 0 & \text{on } \partial\Omega \times (0,T), \end{cases} \tag{2.24}$$

where, again, $(0,T)$ is the *maximal interval* for existence, $T \le \infty$.

The first result, which is well known, says that for d_1, d_2 large, bounded solutions of (2.24) behave roughly like spatially homogeneous solutions, i.e., ODE solutions.

Theorem 2.12. *Suppose that R is a bounded invariant region in $\mathbb{R}^2$ and $\|\nabla f\|_{L^\infty(\Omega)}$, $\|\nabla g\|_{L^\infty(\Omega)} \le M$ on R for some constant $M > 0$. Let the initial value $\{(U(x,0), V(x,0)) \mid x \in \Omega\} \subseteq R$. If $d = \min\{d_1, d_2\} > 2M/\mu_2$, where μ_2 is the first positive eigenvalue of Δ with a homogeneous Neumann boundary condition, then there exists a positive constant $C > 0$ such that for t large*

(i) $\|\nabla_x U(\cdot,t)\|_{L^2(\Omega)} + \|\nabla_x V(\cdot,t)\|_{L^2(\Omega)} \le Ce^{-\sigma t}$;

(ii) $\|U(\cdot,t) - \overline{U}(t)\|_{L^2(\Omega)} + \|V(\cdot,t) - \overline{V}(t)\|_{L^2(\Omega)} \le Ce^{-\sigma t}$,

where $\sigma = d\mu_2 - 2M$, and

$$\overline{U}(t) = \frac{1}{|\Omega|}\int_\Omega U(x,t)\,dx, \quad \overline{V}(t) = \frac{1}{|\Omega|}\int_\Omega V(x,t)\,dx$$

satisfy

$$\begin{cases} \overline{U}_t = f(\overline{U}, \overline{V}) + h_1(t), \\ \overline{V}_t = g(\overline{U}, \overline{V}) + h_2(t) \end{cases}$$

with

$$|h_j(t)| \le Ce^{-\sigma t}.$$

This result is intuitively clear; see, e.g., [Sm, Section 14.D] for a proof.

However, from the modeling point of view, it seems important and mathematically more interesting if 2×2 reaction-diffusion systems could support nontrivial (i.e., *spatially inhomogeneous) stable steady states*. Indeed, in as early as 1952, Turing [T] already used a 2×2 reaction-diffusion system in his attempt to model the "regeneration" phenomenon of *hydra*. The regeneration phenomenon of *hydra*, first discovered by Trembley [Tr] in 1744, is one of the earliest examples in morphogenesis. *Hydra*, an animal of a few millimeters in length, is made up of approximately 100,000 cells of about 15 different types. It consists of a "head" region located at one end along its length. Typical experiments on *hydra* involve removing part of its "head" region and transplanting it to other parts of the body column. Then, a new "head" will form if and only if the transplanted area is sufficiently far from the (old) "head." These observations have led to the assumption of the existence of two chemical substances—a slowly diffusing (short-range) activator and a rapidly diffusing (long-range) inhibitor. In 1952, Turing [T] argued, although diffusion is a smoothing and trivializing process in a system of a single chemical, for systems of two or more chemicals, different diffusion rates could force the uniform steady states to become unstable and lead to nonhomogeneous distributions of such reactants. This is now known as the "diffusion-driven instability." Exploring this idea further, in 1972, Gierer and Meinhardt [GM] proposed the following activator-inhibitor system (already normalized) to model the above regeneration phenomenon of *hydra*:

$$\begin{cases} U_t = d_1 \Delta U - U + \dfrac{U^p}{V^q} & \text{in } \Omega \times (0,T), \\ \tau V_t = d_2 \Delta V - V + \dfrac{U^r}{V^s} & \text{in } \Omega \times (0,T), \\ \partial_\nu U = \partial_\nu V = 0 & \text{on } \partial\Omega \times (0,T), \\ U(x,0) = U_0(x) \geq 0, V(x,0) = V_0(x) > 0 & \text{in } \Omega, \end{cases} \tag{2.25}$$

where the constants τ, d_1, d_2, p, q, r are all positive, $s \geq 0$, and

$$0 < \frac{p-1}{q} < \frac{r}{s+1}. \tag{2.26}$$

Here U represents the density of the slowly diffusing activator which activates both U and V, and V represents the density of the rapidly diffusing inhibitor which suppresses both U and V. Therefore, both U and V are positive, and d_1 is very small while d_2 is very large. The parameter τ here reflects the *response* rate of V versus the change of U.

Condition (2.26) is mathematical, under which one can prove easily that the equilibrium $\overline{U} \equiv 1$ and $\overline{V} \equiv 1$ of the corresponding kinetic system (ODEs)

$$\begin{cases} \overline{U}_t = -\overline{U} + \dfrac{\overline{U}^p}{\overline{V}^q}, \\ \tau \overline{V}_t = -\overline{V} + \dfrac{\overline{U}^r}{\overline{V}^s} \end{cases} \tag{2.27}$$

is stable if $\tau < \frac{s+1}{p-1}$. However, once the diffusion terms are introduced with d_1 small and d_2 large in (2.25), linearized analysis of (2.25) shows that the equilibrium $U \equiv 1$ and $V \equiv 1$ becomes unstable and bifurcations occur; thus "diffusion-driven instability" takes place.

Note that (2.25) does not have variational structure. One way to solve (2.25) is using the "shadow system" approach due to Keener [Ke] in 1978. More precisely, since d_2

is large, we divide the second equation in (2.25) by d_2 and let d_2 tend to ∞. It seems reasonable to expect that, for each fixed t, V tends to a (spatially) harmonic function that must be a constant by the boundary condition. That is, as $d_2 \to \infty$, V tends to a spatially homogeneous function $\xi(t)$. Thus, integrating the second equation in (2.25) over Ω, we reduce (2.25) to the following "shadow system":

$$\begin{cases} U_t = d_1 \Delta U - U + \frac{U^p}{\xi^q} & \text{in } \Omega \times (0,T), \\ \tau \xi_t = -\xi + \dfrac{1}{|\Omega|} \xi^{-s} \displaystyle\int_\Omega U^r(x,t)\,dx & \text{in } (0,T), \\ \partial_\nu U = 0 & \text{on } \partial\Omega \times (0,T), \\ U(x,0) = U_0(x) \geq 0,\ \xi(0) = \xi_0 > 0 & \text{in } \Omega. \end{cases} \tag{2.28}$$

The "shadow system" (2.28) is presumably easier to analyze, and the wish is that properties of solutions to the "shadow system" (2.28) reflect that of solutions to the 2×2 system (2.25), at least for d_2 sufficiently large. To a certain extent, this is true, especially in the study of steady states of (2.25) and (2.28), and we shall address this aspect of the "shadow system" in the next chapter. Here we focus on the dynamic aspect of this "shadow system" approach.

It turns out, somewhat surprisingly, that even as fundamental as the global existence (in time, i.e., $T = \infty$) is concerned, the "shadow system" (2.28) behaves differently from the 2×2 system (2.25), due to a recent result of [LfN].

To describe the situation, we begin with the global existence of (2.25). Although (2.25) was proposed in 1972, and progress had been made since then, the global existence question was largely unsettled until the work of Jiang [J] in 2006.

Theorem 2.13. *If* $\frac{p-1}{r} < 1$, *then every solution of* (2.25) *exists for all time* $t > 0$.

While it has been known [LCQ], [LNT] that for $\frac{p-1}{r} > 1$ even the kinetic system (2.27) blows up at finite time for suitably chosen initial values $(U(0), V(0))$, this only leaves the case $\frac{p-1}{r} = 1$ still open. However, for the "shadow system" (2.28), much less is understood. Obviously, the blow-up result for the kinetic system (2.27) carries over to (2.28). Other than that, we have the following global existence result.

Theorem 2.14. *If* $\frac{p-1}{r} < \frac{2}{n+2}$, *then every solution of* (2.28) *exists for all* $t > 0$.

Compared to Theorem 2.13, this seems quite modest. It seems natural to ask what the *optimal* condition for global existence of (2.28) would be. While we still do not know the answer, the following result of [LfN] does demonstrate a serious gap between the shadow system (2.28) and its original 2×2 Gierer–Meinhardt system (2.25).

Theorem 2.15. *Suppose that* Ω *is the unit ball* $B_1(0)$ *and that* $p = r, \tau = s + 1 - q$, *and* $0 < \frac{p-1}{r} < \frac{q}{s+1} < 1$. *If* $\frac{p-1}{r} > \frac{2}{n}, n \geq 3$, *then* (2.28) *always has finite time blow-up solutions for suitable choices of initial values* U_0 *and* ξ_0.

The range

$$\frac{2}{n} \geq \frac{p-1}{r} \geq \frac{2}{n+2} \tag{2.29}$$

remains open. In proving Theorem 2.15, we first reduce the shadow system (2.28) to a *single nonlocal equation* by considering special initial values for ξ.

Multiplying the equation for ξ in (2.28) by ξ^{s-q}, we have

$$(\xi^{s-q+1})_t + \xi^{s-q} = \frac{1}{|B_1(0)|} \int_{B_1(0)} \frac{U^p}{\xi^q} dx.$$

On the other hand, integrating the equation for U in (2.28), we obtain

$$\overline{U}_t + \overline{U} = \frac{1}{|B_1(0)|} \int_{B_1(0)} \frac{U^p}{\xi^q} dx.$$

Thus

$$(\overline{U} - \xi^{s-q+1})_t + (\overline{U} - \xi^{s-q+1}) = 0.$$

By choosing the initial value $\xi_0^{s-q+1} = \overline{U}_0$, we have $\xi^{s-q+1}(t) \equiv \overline{U}(t)$ and the shadow system (2.28) reduces to the following *single nonlocal equation*:

$$\begin{cases} U_t = d_1 \Delta U - U + \dfrac{U^p}{\overline{U}^{q'}} & \text{in } B_1(0) \times (0,T), \\ \partial_\nu U = 0 & \text{on } \partial B_1(0) \times (0,T), \\ U(x,0) = U_0(x) \geq 0 & \text{in } B_1(0), \end{cases} \tag{2.30}$$

where $q' = \frac{q}{s-q+1}$. We then use ideas in [HY] to construct a sequence of solutions of (2.30) with the blow-up times shrinking to 0.

2.4 Stability Properties of General 2×2 Shadow Systems

2.4.1 General Shadow Systems

Shadow systems are intended to serve as an intermediate step between single equations and 2×2 reaction-diffusion systems. Indeed, the purposes of introducing shadow systems are

(A) *to reflect the behavior of* 2×2 *reaction-diffusion systems when one of the diffusion rates is large;*

(B) *to allow richer dynamics than that of single equations.*

(A) is largely expected to be true, as it has already been witnessed in many important examples. However, we must still proceed with great care, as even for fundamental aspects such as global existence or finite time blow-up, we have seen a remarkable discrepancy in Section 2.3 between the asymptotic behaviors of the 2×2 systems and their shadow systems for the Gierer–Meinhardt system (2.25). For (B), it seems that, indeed, shadow systems do allow richer dynamics, and yet they are not too much more complicated to handle than single equations. We begin by studying the stability properties of steady states of shadow systems.

Recall that for general autonomous single reaction-diffusion equations in convex domains, Theorem 2.6 tells us that *stable steady states must be constants*; in short, "stability implies triviality!" Although counterparts of this result to 2×2 reaction-diffusion systems are yet to be accomplished, it does seem to have a nice extension to general autonomous shadow systems, at least in one space dimension, i.e., when the underlying domain Ω is an interval. In 2001, such a result—"stability implies monotonicity"—was established as a special case of a general theorem which even applies to time-dependent solutions of general autonomous shadow systems [NPY].

In 2008, a result for a general *(not necessarily autonomous)* shadow system, which completely determines *all* the linearized eigenvalues at *any given steady state* (and again contains the "stability implies monotonicity" result as a special case), was obtained by Li, Nakashima, and Ni [LNN].

To describe the result and some of its consequences, we begin with the general 2×2 reaction-diffusion system

$$\begin{cases} U_t = d_1 \Delta U + f(x,U,V) & \text{in } \Omega \times (0,T), \\ \tau V_t = d_2 \Delta V + g(x,U,V) & \text{in } \Omega \times (0,T), \\ \partial_\nu U = \partial_\nu V = 0 & \text{on } \partial\Omega \times (0,T). \end{cases} \tag{2.31}$$

As was described in Section 2.3, the arguments reducing (2.25) to (2.28) work equally well for the general system (2.31). Thus we have formally $V(\cdot,t) \to \xi(t)$ as $d_2 \to \infty$, and (2.31) reduces to

$$\begin{cases} U_t = d_1 \Delta U + f(x,U,\xi) & \text{in } \Omega \times (0,T), \\ \tau \xi_t = \dfrac{1}{|\Omega|} \displaystyle\int_\Omega g(x,U,\xi)dx & \text{for } t \in (0,T), \\ \partial_\nu U = 0 & \text{on } \partial\Omega \times (0,T). \end{cases} \tag{2.32}$$

To study the stability properties of steady states of (2.32), we again use linearized analysis. Let $(U(x),\xi)$ be a steady state of (2.32). Then the existence of an eigenvalue λ with a negative real part of the following linearized problem implies the instability of $(U(x),\xi)$:

$$\begin{cases} L_0\phi + f_\xi(x,U,\xi)\eta + \lambda\phi = 0 & \text{in } \Omega, \\ \dfrac{1}{|\Omega|} \displaystyle\int_\Omega \left[g_U(x,U,\xi)\phi + g_\xi(x,U,\xi)\eta \right] dx + \lambda\eta = 0, & \\ \partial_\nu \phi = 0 & \text{on } \partial\Omega, \end{cases} \tag{2.33}$$

where

$$L_0\phi = d\Delta\phi + f_U(x,U,\xi)\phi.$$

It seems natural to consider the following closely related eigenvalue problem:

$$\begin{cases} L_0\psi + \mu\psi = 0 \quad \text{in } \Omega, \\ \partial_\nu \psi = 0 \quad \text{on } \partial\Omega. \end{cases} \tag{2.34}$$

We denote by $\mu_1 < \mu_2 \le \cdots$ the eigenvalues of (2.34) and by $\psi_1, \psi_2, \ldots$ the corresponding normalized eigenfunctions; i.e.,

$$\langle \psi_i, \psi_j \rangle = \frac{1}{|\Omega|} \int_\Omega \psi_i \psi_j \, dx = \begin{cases} 0 & \text{if } i \ne j, \\ 1 & \text{if } i = j. \end{cases}$$

Our main result here completely determines all the eigenvalues of (2.33).

Theorem 2.16. *The set of all eigenvalues of* (2.33) *consists of the union of three sets* $E_1 \cup E_2 \cup E_3$, *where*

$$E_1 = \{\mu_i \mid \mu_i \text{ is a multiple eigenvalue of } (2.34)\},$$

$$E_2 = \{\mu_j \mid \mu_j \text{ is a simple eigenvalue of } (2.34) \text{ and } a_j b_j = 0\},$$

and

$$E_3 = \left\{ \lambda \neq \mu_k \text{ for all } k \;\middle|\; \tau\lambda = \sum_{k=1}^{\infty} \frac{a_k b_k}{\lambda - \mu_k} - \langle 1, g_\xi(x,U,\xi)\rangle \right\}$$

with $a_k = \langle f_\xi(x,U,\xi), \psi_k\rangle$, $b_k = \langle g_U(x,U,\xi), \psi_k\rangle$.

The following result is a consequence of the above theorem.

Corollary 2.17. *Suppose that* f *is independent of* x *and* (U,ξ) *is a nonconstant steady state of* (2.32). *If* Ω *is convex, then* (U,ξ) *is unstable for all* τ *large.*

In particular, Corollary 2.17 implies that in the autonomous case if ξ responds sufficiently slowly to the change of U, i.e., τ is sufficiently large, then shadow systems can never stabilize a nonconstant steady state in a convex domain. Note that in the autonomous case no condition is assumed for the reaction terms f and g in Corollary 2.17.

Another interesting consequence of Theorem 2.16 is the counterpart of Theorem 2.10 for shadow systems.

Theorem 2.18. *Let* $n \geq 2, \Omega = B_R (B_{R_1} \backslash \overline{B_{R_2}}$, *resp.*), $f(x,U,\xi) = h^2(x)\widehat{f}(U,\xi)$, *with* $h(x) = h(r) > 0$, *where* $r = |x|$, *and let* (U,ξ) *be a nonconstant radially symmetric steady state of* (2.32). *If* h *satisfies*

$$\left[r^{n-1}\left(\frac{1}{r^{n-1}h}\right)' \right]' \geq \frac{n-1}{r^2 h}$$

in Ω, *then* (U,ξ) *is unstable.*

In particular, for general autonomous shadow systems, where $h \equiv 1$, Theorem 2.18 applies and we have the following.

Corollary 2.19. *Let* $n \geq 2$, *and let* Ω *be a ball* $B_R(0)$ *or an annulus* $B_{R_1} \backslash \overline{B_{R_2}}$. *No radially symmetric nontrivial steady states of the shadow system*

$$\begin{cases} U_t = d_1 \Delta U + f(U,\xi) & \text{in } \Omega \times (0,T), \\ \tau\xi_t = \dfrac{1}{|\Omega|}\displaystyle\int_\Omega g(U,\xi)dx & \text{in } (0,T), \\ \partial_\nu U = 0 & \text{on } \partial\Omega \times (0,T) \end{cases} \tag{2.35}$$

can be stable.

For nonautonomous shadow systems, the counterpart of Corollary 2.19 is of course false. A class of such examples is constructed in Proposition 3.3 in [LNN]. (See Example 3.2 in [LNN].)

Incidentally, we ought to remark that for $n = 1$, the "stability implied monotonicity" result established for steady states of general autonomous shadow systems in [NPY] also follows from Theorem 2.16 as a corollary. For, in the autonomous case, if U is not monotone, it must be *k-symmetric* for some integer $k \geq 2$. Then $a_j = b_j = 0$ for all $j = 2, 3, \ldots, k$ and the linearized eigenvalues (all being simple) have the property that $\mu_1 < \cdots < \mu_k < 0$. (See [NPY, Section 2] for details.)

Once again, using the complete classification of all linearized eigenvalues, Theorem 2.16, a counterexample in the nonautonomous case for the "stability implies monotonicity" is included in [LNN, Example 3.1, p. 271].

It should have become clear by now that Theorem 2.16 is also very useful in constructing examples. We will conclude this subsection by mentioning two more examples.

First, a shadow system does not necessarily help stabilizing solutions. In fact, a shadow system could *destabilize* a *stable* steady state of its corresponding single equation.

Proposition 2.20. *Let $\tilde{U}(x)$ be a solution of*

$$\begin{cases} d\Delta U + f(x,U) = 0 & \text{in } \Omega, \\ \partial_\nu U = 0 & \text{on } \partial\Omega, \end{cases}$$

with its linearized eigenvalues $0 < \mu_1 \leq \mu_2 \leq \cdots$; i.e., $\tilde{U}$ is stable. Then, for any fixed C^1 function g and any $\tau > 0$, there exists K_0 such that for $K > K_0$, $(\tilde{U}, \xi_0)$ is an unstable steady state of

$$\begin{cases} U_t = d\Delta U + f(x,U) + \xi - \xi_0 & \text{in } \Omega \times (0,T), \\ \tau \xi_t = \dfrac{1}{|\Omega|} \displaystyle\int_\Omega (g(U) + K\xi)\,dx & \text{in } (0,T), \\ \partial_\nu U = 0 & \text{on } \partial\Omega \times (0,T), \end{cases}$$

where $\xi_0 = -\frac{1}{K}\frac{1}{|\Omega|}\int_\Omega g(\tilde{U})dx$.

See [LNN, Proposition 4.1] for its proof.

Our last example seems curious. Comparing shadow systems to single equations, we may expect that the extra nonlocal equation could help eliminate a one-dimensional unstable manifold at a steady state. Indeed, this is often the case. On the other hand, it also seems intuitively reasonable not to expect shadow systems to be able to stabilize steady states with multidimensional unstable manifolds. This, however, turns out to be false. We refer interested readers to Section 4 of [LNN] for the detailed construction of an autonomous shadow system in one space dimension, i.e., $n = 1$, which stabilizes a steady state of its corresponding single equation with two negative eigenvalues.

In this connection, we ought to remark that, at least for $n = 1$, it is not difficult to see that autonomous shadow systems can never "stabilize" steady states with three-dimensional unstable manifolds, i.e., steady states of the corresponding single equations with at least *three* negative eigenvalues (counting multiplicities).

2.4.2 An Activator-Inhibitor System

In this subsection we wish to return to the original example—the shadow system for the Gierer–Meinhardt system, (2.28) and (2.25).

First, we will describe a stable steady state with a striking pattern, namely, a single spike on the boundary of Ω, for the shadow system (2.28) when d_1 is small:

$$\begin{cases} U_t = \varepsilon^2 \Delta U - U + \dfrac{U^p}{\xi^q} & \text{in } \Omega \times (0,T), \\ \tau \xi_t = -\xi + \dfrac{1}{|\Omega|\xi^s} \displaystyle\int_\Omega U^r\, dx & \text{in } (0,T), \\ \partial_\nu U = 0 & \text{on } \partial\Omega \times (0,T). \end{cases} \tag{2.36}$$

It is not difficult to see that, after rescaling the small parameter $\varepsilon > 0$, a steady state (U,ξ) of (2.36) gives rise to a solution of the following elliptic equation:

$$\begin{cases} \varepsilon^2 \Delta u - u + u^p = 0 & \text{in } \Omega, \\ \partial_\nu u = 0 & \text{on } \partial\Omega. \end{cases} \tag{2.37}$$

Notice that although the original Gierer–Meinhardt system does not have variational structure, equation (2.37) does have an "energy" functional on $H_1(\Omega)$,

$$J_\varepsilon(u) = \frac{1}{2}\int_\Omega (\varepsilon^2|\nabla u|^2 + u^2)\, dx - \frac{1}{p+1}\int_\Omega u_+^p dx, \tag{2.38}$$

where $u_+ = \max\{u, 0\}$. *Here we assume that* $1 < p < \frac{n+2}{n-2}$. Although "energy" functional $J_\varepsilon(u)$ is neither bounded from above nor bounded from below, it does have a *positive* minimum on the set of all positive solutions of (2.37); in fact, the following constrained minimum is assumed [NT2]:

$$J_\varepsilon(u_\varepsilon) = \min_{u \in \mathfrak{W}} J_\varepsilon(u), \tag{2.39}$$

where

$$\mathfrak{W} = \{u \in H_1(\Omega) \,|\, u \geq 0, u \neq 0, I_\varepsilon(u) = 0\} \tag{2.40}$$

with

$$I_\varepsilon(u) = \int_\Omega (\varepsilon^2|\nabla u|^2 + u^2) - \int_\Omega u^{p+1}. \tag{2.41}$$

(Observe that $\mathfrak{W}$ contains *all positive solutions* of (2.37).) It turns out that *the minimizer* $u_\varepsilon \in \mathfrak{W}$ *is a positive solution* of (2.37) and therefore will be referred to as a *least-energy solution of* (2.37). Some of the striking features of u_ε are summarized in the following theorem, which is due to Ni and Takagi from about 20 years ago [NT2, NT3].

Theorem 2.21 (see [NT2, NT3]). *Suppose that* $1 < p < \frac{n+2}{n-2}$. *Then for every* $\varepsilon > 0$ *small* (2.37) *has a least-energy solution* u_ε *with the following properties:*

(i) $u_\varepsilon > 0$ *on* $\overline{\Omega}$ *and* u_ε *has a unique local (thus global) maximum point* P_ε *on* $\overline{\Omega}$. *Furthermore,* $P_\varepsilon \in \partial\Omega$ *and*

$$H(P_\varepsilon) \to \max_{P \in \partial\Omega} H(P) \quad \text{as } \varepsilon \to 0,$$

where H *denotes the mean curvature of* $\partial\Omega$. *In other words,* P_ε *is located near the "most curved" part of the boundary* $\partial\Omega$.

(ii) $u_\varepsilon(P_\varepsilon) \to w(0)$ *as* $\varepsilon \to 0$, *where* w *is the unique positive solution of the following problem on the entire* $\mathbb{R}^n$:

$$\begin{cases} \Delta w - w + w^p = 0 & \text{in } \mathbb{R}^n, \\ w > 0 & \text{in } \mathbb{R}^n, \\ w(0) = \max_{\mathbb{R}^n} w, \quad w \to 0, & \text{at } \infty. \end{cases} \tag{2.42}$$

In fact, the profile of the "spike" is also determined to the first order of ε in [NT3]. We will give a more detailed description of u_ε in Chapter 3, together with much further development in this important direction.

With this least-energy solution u_ε of (2.37), we obtain easily a corresponding steady state solution $(U_\varepsilon, \xi_\varepsilon)$ for the shadow system (2.36),

$$U_\varepsilon = \xi_\varepsilon^{\frac{q}{p-1}} u_\varepsilon, \quad \xi_\varepsilon = \left(\frac{1}{|\Omega|} \int_\Omega u_\varepsilon \right)^{-\frac{1}{\alpha}}, \tag{2.43}$$

where

$$\alpha = \frac{qr}{p-1} - (s+1) > 0.$$

Various stability and instability properties of $(U_\varepsilon, \xi_\varepsilon)$ are studied in [NTY2]. Among other things, the following stability result holds.

Theorem 2.22. *Suppose that* $r = p+1$ *and* $1 < p < \frac{n+2}{n-2}$. *Then, for* $\alpha \notin \mathcal{C}$, *where* $\mathcal{C}$ *is a certain finite (possibly empty) subset of* $(0, \frac{qr}{p-1} - 1)$, *we have that all the linearized eigenvalues of* (2.36) *at* $(U_\varepsilon, \xi_\varepsilon)$ *are contained in the set*

$$\{\lambda \in \mathbb{C} \,|\, Re\, \lambda < 0 \text{ or } \lambda = 0\},$$

provided that $\tau > 0$ *is sufficiently small. In particular,*

(i) *if the maximum point* P_ε *approaches a nondegenerate maximum point of the boundary mean curvature function* $H(P)$, *then* 0 *is not in the linearized spectrum and* $(U_\varepsilon, \xi_\varepsilon)$ *is asymptotically stable;*

(ii) *if* Ω *is a ball or an annulus in* $\mathbb{R}^2$, *then* 0 *is a linearized eigenvalue, but* $(U_\varepsilon, \xi_\varepsilon)$ *is still asymptotically stable.*

(See Theorem B and Proposition 4.2 in [NTY2] for the detailed proofs.) Thus, the shadow system (2.36) does support stable steady states with a single spike.

On the other hand, in 2007, Miyamoto [My] proved the following theorem, complementing the above result.

Theorem 2.23. *Suppose that* $r = p+1$ *and* $\Omega = B_R(0)$ *in* $\mathbb{R}^2$. *If* (U, ξ) *is a stable steady state of* (2.36), *then the maximum (minimum) of* U *is attained at exactly one point on* $\partial\Omega$, *and* U *has no critical point in (the interior of)* Ω.

Note that in the theorem above, Miyamoto *does not* need to assume that the diffusion rate ε^2 of U is small.

Combining Theorems 2.22 and 2.23, we see that in case Ω is a ball in $\mathbb{R}^2$ and $r = p+1$, *only single boundary spike-layer steady states for* (2.36) *could be stable, and under appropriate conditions, those single boundary spike-layer steady states are indeed stable!*

Chapter 3

Qualitative Properties of Steady States of Reaction-Diffusion Equations and Systems

Steady states often play important roles in the study of parabolic equations and systems. Although systematic studies only essentially started in the late 1970s, this area has experienced a vast development in the last 30 years. One of the main purposes of this chapter is to introduce some of the relevant results, especially those simple, fundamental, and related to the diffusion rates and/or different boundary conditions, to the interested readers.

Systematic studies of qualitative properties of solutions to general nonlinear elliptic equations or systems essentially began in the late 1970s, although some nonlinear elliptic equations (such as the Lane–Emden equation in astrophysics [Ch]) actually go back to the 19th century. It should be noted, however, that earlier works in this direction on linear elliptic equations, such as symmetrization or nodal properties of eigenfunctions, have had their consequences in nonlinear equations. (See, e.g., [PS], [Ch].)

Symmetry remains an important topic in modern theory of nonlinear PDEs. In particular, it is now understood how different boundary conditions may influence the symmetry properties of positive solutions in domains with symmetries. First, solutions of boundary-value problems are very different from solutions on entire space. Moreover, solutions to Neumann boundary-value problems exhibit behavior drastically different from their Dirichlet counterparts. For instance, it is known [GNN1] that *all positive solutions of the Dirichlet problem*

$$\begin{cases} d\Delta u + f(u) = 0 & \text{in } B_R(0), \\ u = 0 & \text{on } \partial B_R(0), \end{cases}$$

where f is a locally Lipschitz continuous function and $B_R(0)$ is the ball of radius R centered at the origin 0, *must be radially symmetric, regardless of the diffusion rate* d. On the other hand, it was proved in [NT2] that for the diffusion rate d *sufficiently small the Neumann problem*

$$\begin{cases} d\Delta u - u + u^p = 0 & \text{in } B_R(0), \\ \partial_\nu u = 0 & \text{on } \partial B_R(0), \end{cases}$$

where $1 < p < \frac{n+2}{n-2}$ $(= \infty$ if $n = 2)$, *possesses a positive solution* u_d *with a unique maximum point located on the boundary* $\partial B_R(0)$. Thus, this solution u_d cannot possibly be radially symmetric. In fact, *for* d *large,* $u \equiv 1$ *is the only positive solution, and the num-*

ber of positive nonradial solutions of the Neumann problem above tends to ∞ *as* d *tends to* 0. Furthermore, while it has been known for decades that symmetrization reduces the "energy" of positive solutions for Dirichlet problems, it can be shown that symmetrization actually increases the "energy" of the solution u_d above. (Here, by "energy" we mean the variational integral

$$\int_{B_R(0)} \left[\frac{1}{2}\left(d\,|\nabla u|^2 + u^2 \right) - \frac{1}{p+1} u^{p+1} \right].$$

Note that, since symmetrization does not alter integrals involving u, only the Dirichlet integral

$$\int_{B_R(0)} |\nabla u|^2$$

gets changed after symmetrization.) In other words, the most "stable" solutions to the Neumann problem above *must* not be radially symmetric—a remarkable difference between Neumann and Dirichlet boundary conditions. In fact, solutions to Neumann problems also possess some restricted symmetry properties—they seem to be more subtle. (See Section 3.3.) Generally speaking, Dirichlet boundary conditions are far more rigid and imposing than Neumann boundary conditions, as is already indicated by the above discussions. This is also true for general bounded smooth domains Ω in $\mathbb{R}^n$.

Symmetry properties of solutions to elliptic equations on entire space (or unbounded domains) clearly require appropriate conditions at ∞. It seems that the simplest result in this direction is that *all positive solutions of the problem*

$$\begin{cases} \Delta u + f(u) = 0 & \textit{in } \mathbb{R}^n, \\ u \to 0 & \textit{at } \infty \end{cases}$$

must be radially symmetric (up to a translation), provided that $f'(0) < 0$. (See [GNN2], [LiN], or Theorem 3.15 below.) The case $f'(0) = 0$ turns out to be far more complicated. Roughly speaking, to guarantee radial symmetry in this case, additional hypotheses on suitable decay of solutions are needed unless $f'(s) \leq 0$ for all sufficiently small $s > 0$. (See Theorem 3.15.) Symmetries and related properties, such as monotonicity, are discussed in Section 3.3.

In a different but very important direction, significant progress has been made in the past 20 years in understanding the "shape" of solutions, in particular, the "concentration" behavior of solutions to nonlinear elliptic equations and systems. More precisely, positive solutions concentrating near isolated points, i.e., *spike-layer solutions* (or, single- and multipeak solutions), and the locations of these points (determined by the geometry of the underlying domains) have been obtained for both Dirichlet and Neumann boundary-value problems. For instance, as was mentioned in the previous chapter, for ε small, a "least-energy solution" of the Neumann problem

$$\begin{cases} \varepsilon^2 \Delta u - u + u^p = 0 & \text{in } \Omega, \\ u > 0 & \text{in } \Omega, \\ \partial_\nu u = 0 & \text{on } \partial\Omega, \end{cases} \tag{3.1}$$

where $1 < p < \frac{n+2}{n-2}$ $(= \infty$ if $n = 2)$, *must have its only (local and thus global) maximum point (in* $\overline{\Omega}$*) located on* $\partial\Omega$ *and near the most "curved" part of* $\partial\Omega$. (See [NT2, NT3] or

Theorem 3.3. Here, the "curvedness" is measured by the mean curvature of $\partial\Omega$.) On the other hand, a "least-energy solution" of the Dirichlet problem

$$\begin{cases} \varepsilon^2 \Delta u - u + u^p = 0 & \text{in } \Omega, \\ u > 0 & \text{in } \Omega, \\ u = 0 & \text{on } \partial\Omega, \end{cases} \tag{3.2}$$

where $1 < p < \frac{n+2}{n-2}$ ($= \infty$ if $n = 2$), *must have its only (local and thus global) maximum point (in $\overline{\Omega}$) located near a "center" of the domain* Ω. (See [NWe] or Theorem 3.4.) Here a "center" is defined as a point in Ω which is *most distant from* $\partial\Omega$.) Furthermore, the "profiles" of these least-energy solutions for both (3.1) and (3.2) have been determined in [NT2, NT3] and [NWe]. There has been a huge amount of literature on those spike-layer solutions published since the papers [NT2, NT3] first appeared in the early 1990s, and many interesting and excellent results have been obtained. For example, the locations of multiple interior peaks to a solution of (3.1), for ε small, are determined by the "sphere-packing" property of the domain Ω. (See [GW1] or Theorem 3.5.) Those solutions often represent pattern formation in various branches of sciences. In Sections 3.1 and 3.2, we shall describe the recent progress in this direction as well as some models leading to those solutions. (We will, for instance, include the Gierer–Meinhardt system mentioned in the previous chapter.) Furthermore, positive solutions concentrating on multidimensional subsets (instead of isolated points which are zero dimensional) of the underlying domains will also be discussed in Section 3.1, although advance in this direction has been rather limited so far.

The "shape" of solutions of elliptic equations or systems turns out to be closely related to the stability properties of those solutions. As we have seen, stability properties are crucial to our understanding of the entire dynamics of the original evolution problems. Roughly speaking, the "general principle" here seems to be that *the "simpler the shape" of a solution, the "more stable" it tends to be.* (See Chapter 2 for more detailed discussions.) Therefore, it seems that efforts in understanding the "shape" of steady states are necessary, and, hopefully, readers will agree that efforts in this direction are rewarding.

Finally, we remark that we shall consider only "autonomous" equations and systems (i.e., no explicit dependence in spatial variables appearing in the equations and systems) in this chapter. Remarks concerning various generalizations are included at appropriate places; see, e.g., Subsections 3.1.4 and 3.3.6. However, it is important to point out that spatial heterogeneity, when interacting with diffusions, could produce extremely interesting phenomena. We shall discuss those in Chapters 4 and 5.

3.1 Concentrations of Solutions: Single Equations

One of the greatest advances in the theory of PDEs is the recent progress on the studies of concentration behaviors of solutions to elliptic equations and systems. It is remarkable to see that similar, and in many cases independent, results have been obtained concerning these striking behaviors in various models from different areas of science. These include activator-inhibitor systems in modeling the regeneration phenomenon of *hydra*, Ginzburg–Landau systems in superconductivity, nonlinear Schrödinger equations, the Gray–Scott model, the Lotka–Volterra competition system with cross-diffusions, and others. In this and the next sections, we shall include descriptions of some of these systems from their backgrounds to the significance of the mathematical results obtained. Comparisons of results under different boundary conditions also will be made to illustrate the importance of boundary effects on the behaviors of solutions.

3.1.1 Spike-Layer Solutions in Elliptic Boundary-Value Problems

We have indicated in the introduction that the Neumann boundary condition is far less restrictive than the Dirichlet boundary condition. Consequently, Neumann boundary-value problems tend to allow more solutions with more interesting behaviors than their Dirichlet counterparts. However, it is interesting to note that systematic studies of nonlinear Neumann problems seem to have a much shorter history.

Studies of nonlinear Neumann problems are often motivated by models in pattern formation in physical or biological sciences. One of the more well-known examples is Turing's "diffusion-driven instability," which led to the Gierer–Meinhardt system (2.25) in modeling the regeneration phenomenon in *hydra*, described in Section 2.3. In particular, one is led to seeking "spike-layer" solutions of (3.1). (See (2.37) in Section 2.4.) Here we shall give a more systematic introduction of (3.1), along with (3.2), for comparison purposes.

Recall briefly that in 1952, Turing [T] proposed the ingenious idea of "diffusion-driven instability," which says that for systems of two or more chemicals, different diffusion rates could force the uniform steady states to become unstable and lead to nonhomogeneous distributions of such reactants. Exploring this idea further, in 1972, Gierer and Meinhardt [GM] proposed the following activator-inhibitor system (already normalized) to model the regeneration phenomenon of *hydra*:

$$\begin{cases} U_t = d_1 \Delta U - U + \dfrac{U^p}{V^q} & \text{in } \Omega \times (0,T), \\ \tau V_t = d_2 \Delta V - V + \dfrac{U^r}{V^s} & \text{in } \Omega \times (0,T), \\ \partial_\nu U = \partial_\nu V = 0 & \text{on } \partial\Omega \times (0,T), \end{cases} \tag{3.3}$$

where the constants τ, d_1, d_2, p, q, r are all positive, $s \geq 0$, and

$$0 < \frac{p-1}{q} < \frac{r}{s+1}. \tag{3.4}$$

(See Section 2.3 for a more detailed description.) Here U represents the density of the slowly diffusing activator which activates both U and V, and V represents the density of the rapidly diffusing inhibitor which suppresses both U and V. Therefore, both U and V are positive, and d_1 is very small while d_2 is very large. The parameter τ here reflects the *response* rate of V versus the change of U.

One way to solve (3.3) is using the "shadow system" approach. More precisely, since d_2 is large, we divide the second equation in (3.3) by d_2 and let d_2 tend to ∞. It seems reasonable to expect that, for each fixed t, V tends to a (spatially) harmonic function that must be a constant by the boundary condition. That is, as $d_2 \to \infty$, V tends to a spatially homogeneous function $\xi(t)$. Thus, integrating the second equation in (3.3) over Ω, we reduce (3.3) to the following "shadow system":

$$\begin{cases} U_t = d_1 \Delta U - U + \dfrac{U^p}{\xi^q} & \text{in } \Omega \times (0,T), \\ \tau \xi_t = -\xi + \dfrac{1}{|\Omega|}\xi^{-s} \displaystyle\int_\Omega U^r(x,t)\,dx & \text{in } (0,T), \\ \partial_\nu U = 0 & \text{on } \partial\Omega \times (0,T), \end{cases} \tag{3.5}$$

where $|\Omega|$ denotes the measure of Ω. Although the above reduction can be verified rigorously in some cases [NT1], we must point out that it is more important to solve (3.3) via solutions of (3.5). It turns out that the steady states of (3.5) and their stability properties are closely related to those of the original system (3.3) and that the study of the steady states of (3.5) essentially reduces to that of the following single equation (by a suitable scaling argument):

$$\begin{cases} \varepsilon^2 \Delta u - u + u^p = 0 & \text{in } \Omega, \\ u > 0 & \text{in } \Omega, \\ \partial_\nu u = 0 & \text{on } \partial\Omega. \end{cases} \tag{3.1}$$

In the case $n = 1$, a lot of work has been done by Takagi [Ta]. For $n \geq 2$, the situation becomes far more interesting. The pioneering works [NT1, NT2, NT3], [LNT] produced a single-peak spike-layer solution u_ε of (3.1) in 1993. Furthermore, steady states of the shadow system (3.5) as well as the original system (3.3) have been constructed from u_ε—at least for small d_1 and large d_2—and their stability properties have been investigated [NT4], [NTY1, NTY2].

It seems illuminating to "solve" (3.1) as well as its Dirichlet counterpart side by side,

$$\begin{cases} \varepsilon^2 \Delta u - u + u^p = 0 & \text{in } \Omega, \\ u > 0 & \text{in } \Omega, \\ u = 0 & \text{on } \partial\Omega, \end{cases} \tag{3.2}$$

and compare the qualitative properties of the solutions.

For (3.1), first and foremost, we mention that the diffusion coefficient ε^2 is important in the existence as well as the "shapes" of its nontrivial solutions.

Theorem 3.1. *Suppose that* $1 < p < \frac{n+2}{n-2}$ ($= \infty$ *when* $n = 2$). *Then* $u \equiv 1$ *is the only solution of* (3.1) *for* ε *large.*

This result is due to [LNT]. The conclusion is intuitively clear once an *a priori bound independent of the diffusion rate* ε, for the L^∞ norms of the solutions, is obtained.

As we will see later in this section, as ε decreases to 0, the number of solutions of (3.1) tends to ∞ and the pattern of the solutions also becomes more and more sophisticated.

On the other hand, the story for the Dirichlet problem (3.2) is very different. It is well known that the following result holds.

Theorem 3.2. *Suppose that* $1 < p < \frac{n+2}{n-2}$ ($= \infty$ *when* $n = 2$). *Then* (3.2) *always possesses a solution, regardless of the size of* ε. *Furthermore, for* $p \geq \frac{n+2}{n-2}$, (3.2) *has no solution if* Ω *is star-shaped.*

The existence part of Theorem 3.2 can be obtained by a standard variational approach—either constrained minimization or the Mountain-Pass lemma. The nonexistence for star-shaped domains follows from the well-known Rellich–Pohozaev identity.

Two more remarks for (3.2) are in order. First, the number of solutions of (3.2) seems small in general and does not seem to depend on the diffusion rate ε^2. For instance, when Ω is a ball, (3.2) has a *unique* solution (for $1 < p < \frac{n+2}{n-2}$) for every $\varepsilon > 0$. Second, the "shapes" of solutions of (3.2) also seem rather simple, as we shall see next.

One common feature of (3.1) and (3.2) is that both problems exhibit "concentration" phenomena for ε small. More precisely, *both* (3.1) *and* (3.2) *possess single-peak spike-layer solutions for* ε *small.*

We shall first describe how the existence of a *single-peak* spike-layer solution is established and then discuss the location and the profile of this single peak. Since the equation in (3.1) and (3.2) is "autonomous" (i.e., no explicit spatial dependence in the equation), the location of the spike must be determined by the geometry of Ω. We would like to call the reader's attention to see exactly how the geometry of Ω enters the picture in each of the problems (3.1) and (3.2) separately and to compare the effects of *different* boundary conditions on the location of the peak.

For ε small, (3.1) and (3.2) are singular perturbation problems. However, the traditional method in applied mathematics, using inner and outer expansions, simply does not apply here, because a spike-layer solution of (3.1) or (3.2) is exponentially small away from its peaks.

In the rest of the section, we will always assume that $1 < p < \frac{n+2}{n-2}$ if $n \geq 3$, and $1 < p < \infty$ if $n = 1,2$. We first define the "energy" functional in $H_1(\Omega)$

$$J_{\varepsilon,N}(u) = \frac{1}{2}\int_\Omega \left(\varepsilon^2 |\nabla u|^2 + u^2\right) - \frac{1}{p+1}\int_\Omega u_+^{p+1}, \tag{3.6}$$

where $u_+ = \max\{u, 0\}$. It is standard to check that a critical point corresponding to a *positive* critical value of $J_{\varepsilon,N}$ is a classical solution of (3.1). Similarly, we define the "energy" functional in $H_1^0(\Omega)$

$$J_{\varepsilon,D}(u) = \frac{1}{2}\int_\Omega \left(\varepsilon^2 |\nabla u|^2 + u^2\right) - \frac{1}{p+1}\int_\Omega u_+^{p+1} \tag{3.7}$$

and observe that a critical point corresponding to a *positive* critical value of $J_{\varepsilon,D}$ is a classical solution of (3.2). Our first step appears to be nothing unusual; namely, we shall use the well-known Mountain-Pass lemma to guarantee that each of $J_{\varepsilon,N}$ and $J_{\varepsilon,D}$ has a positive critical value. However, in order to use this variational formulation to obtain useful information later, our formulation of the Mountain-Pass lemma follows Ding and Ni [DN] and deviates from the usual one. (See Subsection 3.1.4.) More precisely, setting

$$c_{\varepsilon,N} = \inf\left\{\max_{t\geq 0} J_{\varepsilon,N}(tv) \,\middle|\, v \geq 0, \not\equiv 0 \text{ in } H_1(\Omega)\right\} \tag{3.8}$$

and

$$c_{\varepsilon,D} = \inf\left\{\max_{t\geq 0} J_{\varepsilon,D}(tv) \,\middle|\, v \geq 0, \not\equiv 0 \text{ in } H_1^0(\Omega)\right\}, \tag{3.9}$$

we show that $c_{\varepsilon,N}$ is a positive critical value of $J_{\varepsilon,N}$, thus giving rise to a solution $u_{\varepsilon,N}$ of (3.1); and similarly we show that $c_{\varepsilon,D}$ is a positive critical value of $J_{\varepsilon,D}$, thus giving rise to a solution $u_{\varepsilon,D}$ of (3.2). Our main task here is to prove that both $u_{\varepsilon,N}$ and $u_{\varepsilon,D}$ exhibit a single spike-layer behavior, and we are going to determine the locations as well as the profiles of the spike layer of $u_{\varepsilon,N}$ and $u_{\varepsilon,D}$.

Roughly speaking, both $u_{\varepsilon,N}$ and $u_{\varepsilon,D}$ can have one "peak" (i.e., a local maximum point in $\overline{\Omega}$), denoted by $P_{\varepsilon,N}$ and $P_{\varepsilon,D}$, respectively, and must tend to 0 everywhere else. Moreover, $P_{\varepsilon,N}$ must lie on the boundary $\partial\Omega$ and tend to the "most-curved" part of $\partial\Omega$, while $P_{\varepsilon,D}$ must tend to the "most-centered" part of Ω in the interior as ε tends to 0. As for the profiles of $u_{\varepsilon,N}$ and $u_{\varepsilon,D}$, again, roughly speaking, $u_{\varepsilon,D}$ is approximately a "scaled" version of w near $P_{\varepsilon,D}$, where w is the unique solution of

$$\begin{cases} \Delta w - w + w^p = 0 & \text{in } \mathbb{R}^n, \\ w > 0 \quad \text{in } \mathbb{R}^n, & w \to 0 \text{ at } \infty, \\ w(0) = \max\, w, \end{cases} \tag{3.10}$$

while $u_{\varepsilon,N}$ is approximately a "scaled" and "deformed" version of "half" of w. To make those descriptions precise, we start with $u_{\varepsilon,N}$.

Theorem 3.3. *For each ε sufficiently small, the solution $u_{\varepsilon,N}$ has exactly one local (thus global) maximum point in $\bar{\Omega}$ and it is achieved at exactly one point $P_{\varepsilon,N}$ in $\bar{\Omega}$. Moreover, $u_{\varepsilon,N}$ has the following properties:*

(i) *As $\varepsilon \to 0$ the translated solution $u_\varepsilon(\cdot + P_{\varepsilon,N}) \to 0$ except at 0, and $u_{\varepsilon,N}(P_\varepsilon, N) \to w(0)$, where w is the unique solution of* (3.10).

(ii) *$P_{\varepsilon,N} \in \partial\Omega$ and $H(P_{\varepsilon,N}) \to \max_{P\in\partial\Omega} H(P)$ as $\varepsilon \to 0$, where H denotes the mean curvature of $\partial\Omega$.*

(iii) *Through rotation and translation we may suppose that $P_{\varepsilon,N}$ is the origin and near $P_{\varepsilon,N}$ the boundary $\partial\Omega = \{(x', x_n) \,|\, x_n = \psi_\varepsilon(x')\}$ and $\Omega = \{(x', x_n) \,|\, x_n > \psi_\varepsilon(x')\}$, where $x' = (x_1, \ldots, x_{n-1})$, and $\psi_\varepsilon(0) = 0$, $\nabla\psi_\varepsilon(0) = 0$. Then the diffeomorphism $x = \Phi_\varepsilon(\widetilde{x}) = (\Phi_{\varepsilon,1}(\widetilde{x}), \ldots, \Phi_{\varepsilon,n}(\widetilde{x}))$ defined by*

$$\Phi_{\varepsilon,j}(\widetilde{x}) = \begin{cases} \widetilde{x}_j - \widetilde{x}_n \frac{\partial\psi_\varepsilon}{\partial x_j}(\widetilde{x}') & \text{for } j = 1, \ldots, n-1, \\ \widetilde{x}_n + \psi_\varepsilon(\widetilde{x}') & \text{for } j = n \end{cases}$$

flattens the boundary $\partial\Omega$ near $P_{\varepsilon,N}$, and

$$u_{\varepsilon,N}(\Phi_\varepsilon(\varepsilon y)) = w(y) + \varepsilon\phi(y) + o(\varepsilon), \tag{3.11}$$

where $\tilde{x} = \varepsilon y$ and ϕ is the unique solution of

$$\begin{cases} \Delta\phi - \phi + p w^{p-1}\phi \\ \quad + 2|y_n| \sum_{i,j=1}^{n-1} \psi_{\varepsilon,ij} \frac{\partial^2 w}{\partial y_i \partial y_j} - \alpha_\varepsilon (\operatorname{sgn} y_n) \frac{\partial w}{\partial y_n} = 0 \;\; \text{in } \mathbb{R}^n, \\ \phi(y) \to 0 \text{ as } \;\; y \to \infty, \;\; \text{and } \int_{\mathbb{R}^n} \phi \frac{\partial w}{\partial y_j} = 0 \text{ for } j = 1, \ldots, n, \end{cases} \tag{3.12}$$

with $\psi_{\varepsilon,ij} = \frac{\partial^2\psi_\varepsilon}{\partial x_i \partial x_j}(0), \alpha_\varepsilon = \Delta\psi_\varepsilon(0)$.

Note that (3.11) gives the profile of $u_{\varepsilon,N}$ up to the second order, and it can be proved that ϕ actually decays exponentially near ∞. The detailed proof of Theorem 3.3 may be found in [NT2, NT3].

We now turn to the Dirichlet case. To describe our results, first we need to introduce some notation. For a bounded smooth domain $\widetilde{\Omega}$ in $\mathbb{R}^n$, we let $\mathcal{P}_{\widetilde{\Omega}}w$ be the solution of the linear problem

$$\begin{cases} \Delta v - v + w^p = 0 & \text{in } \widetilde{\Omega}, \\ v = 0 & \text{on } \partial\widetilde{\Omega}, \end{cases} \tag{3.13}$$

where w is the unique solution of (3.10). Now, set

$$z = \frac{x - P_{\varepsilon,D}}{\varepsilon}$$

and $\Omega_\varepsilon = \{z \in \mathbb{R}^n \,\big|\, x = P_{\varepsilon,D} + \varepsilon z \in \Omega\}$, where $P_{\varepsilon,D}$ (to be determined) is the unique peak of $u_{\varepsilon,D}$ as is stated in Theorem 3.4. Since eventually we will show the scaled version of $\mathcal{P}_{\Omega_\varepsilon}w$ is a very good approximation of u_ε, we need to study the difference between w and $\mathcal{P}_{\Omega_\varepsilon}w$, that is, the function $\varphi_\varepsilon \equiv w - \mathcal{P}_{\Omega_\varepsilon}w$, which satisfies

$$\begin{cases} \Delta\varphi_\varepsilon - \varphi_\varepsilon = 0 & \text{in } \Omega_\varepsilon, \\ \varphi_\varepsilon = w & \text{on } \partial\Omega_\varepsilon. \end{cases} \tag{3.14}$$

The quantity φ_ε is extremely small, and it turns out that the "correct" order of the difference $\omega - \mathcal{P}_{\Omega_\varepsilon}\omega$ (for our purposes) is the logarithmic of $\varphi_\varepsilon^{-\varepsilon}$, i.e., the function $\delta_\varepsilon(x) = -\varepsilon\log\varphi_\varepsilon(z)$, which satisfies a nonlinear equation instead:

$$\begin{cases} \varepsilon\Delta\delta_\varepsilon - |\nabla\delta_\varepsilon|^2 + 1 = 0 & \text{in } \Omega, \\ \delta_\varepsilon(x) = -\varepsilon\log w\left(\frac{x - P_{\varepsilon,D}}{\varepsilon}\right) & \text{on } \partial\Omega. \end{cases} \tag{3.15}$$

Finally, we enlarge φ_ε to $V_\varepsilon(z) = e^{\frac{1}{\varepsilon}\delta_\varepsilon(P_{\varepsilon,D})}\varphi_\varepsilon(z)$. It is clear that V_ε satisfies

$$\begin{cases} \Delta V_\varepsilon - V_\varepsilon = 0 & \text{in } \Omega, \\ V_\varepsilon(0) = 1. \end{cases}$$

We are now ready to state our main results for the Dirichlet problem (3.2).

Theorem 3.4. *For each ε sufficiently small, the solution $u_{\varepsilon,D}$ has exactly one local (thus global) maximum in Ω and it is achieved at exactly one point $P_{\varepsilon,D}$ in Ω. Moreover, $u_{\varepsilon,D}$ has the following properties:*

(i) *As $\varepsilon \to 0$ the translated solution $u_{\varepsilon,D}(\cdot + P_{\varepsilon,D}) \to 0$ except at 0, and $u_{\varepsilon,D}(P_{\varepsilon,D}) \to w(0)$, where w is the unique solution of* (3.10).

(ii) *$d(P_{\varepsilon,D}, \partial\Omega) \to \max_{P\in\Omega} d(P, \partial\Omega)$ as $\varepsilon \to 0$, where d denotes the usual distance function.*

(iii) *For every sequence $\varepsilon_k \to 0$, there is a subsequence $\varepsilon_{k_i} \to 0$ such that for $\varepsilon = \varepsilon_{k_i}$ it holds that*

$$u_{\varepsilon,D}(x) = \mathcal{P}_{\Omega_\varepsilon}w(z) + e^{-\frac{1}{\varepsilon}\delta_\varepsilon(P_{\varepsilon,D})}\phi_\varepsilon(z) + o\left(e^{-\frac{1}{\varepsilon}\delta_\varepsilon(P_{\varepsilon,D})}\right), \tag{3.16}$$

where $\delta_\varepsilon(P_\varepsilon, D) \to 2\max_{P\in\Omega} d(P,\partial\Omega)$, *and* $\left\|e^{-\mu|z|}(\phi_\varepsilon - \phi_0)\right\|_{L^\infty(\mathbb{R}^n)} \to 0$ *with* $1 > \mu > \max\{0, 2-p\}, \phi_0$ *being a solution of*

$$(\Delta - 1 + pw^{p-1})\phi_0 = pw^{p-1}V_0 \quad \text{in } \mathbb{R}^n,$$

and V_0 *being the pointwise limit of* V_ε, $\varepsilon = \varepsilon_{k_i}$.

Several remarks are in order. First, comparing Theorems 3.3 and 3.4, we see that part (i) of Theorems 3.3 and 3.4, respectively, shows that each of the solutions $u_{\varepsilon,N}$ and $u_{\varepsilon,D}$ possesses a single-peak spike-layer structure, and part (ii) of Theorems 3.3 and 3.4, respectively, also, locates the peak of $u_{\varepsilon,N}$ and of $u_{\varepsilon,D}$. It is interesting to note that although intuitively, by the exponential decay of w, a scaled w (i.e., $w(\frac{x-P_{\varepsilon,D}}{\varepsilon})$) which is truncated near $\partial\Omega$ seems to be an excellent approximation for $u_{\varepsilon,D}$, part (iii) of Theorem 3.4 indicates that the function $\mathcal{P}_{\Omega_\varepsilon} w(\frac{x-P_{\varepsilon,D}}{\varepsilon})$ is actually a *better* approximation for $u_{\varepsilon,D}$. This is quite delicate since the error terms induced by these two approximations are both of exponentially small order and are very close. In fact, this observation turns out to be crucial in pushing our method through for the Dirichlet case (3.2).

We now describe our method of proofs. In both cases (3.1) and (3.2), the most important idea is to obtain an estimate for $c_{\varepsilon,N}$ and $c_{\varepsilon,D}$, respectively, which is sufficiently accurate to reflect the influence of the geometry of the domain Ω. More precisely, both the zeroth-order approximation for $c_{\varepsilon,N}$ and that for $c_{\varepsilon,D}$ depend only on the unique solution w (and its energy) of (3.10). The geometry of the domain Ω, namely, the boundary mean curvature $H(P_{\varepsilon,N})$ at $P_{\varepsilon,N}$ in the Neumann case and the distance $d(P_{\varepsilon,D}, \partial\Omega)$ in the Dirichlet case, enters the first-order approximation of $c_{\varepsilon,N}$ and $c_{\varepsilon,D}$. To be explicit, we have

$$c_{\varepsilon,N} = \varepsilon^n \left\{ \frac{1}{2} J(w) - (n-1)\gamma_N H(P_{\varepsilon,N})\varepsilon + o(\varepsilon) \right\} \tag{3.17}$$

and

$$c_{\varepsilon,D} = \varepsilon^n \left\{ J(w) + e^{-\frac{1}{\varepsilon}\delta_\varepsilon(P_{\varepsilon,D})}\gamma_D + o\left(e^{-\frac{1}{\varepsilon}\delta_\varepsilon(P_{\varepsilon,D})}\right) \right\}, \tag{3.18}$$

where γ_N and γ_D are positive constants independent of ε, and

$$J(w) = \frac{1}{2}\int_{\mathbb{R}^n} \left(|\nabla w|^2 + w^2\right) - \frac{1}{p+1}\int_{\mathbb{R}^n} w^{p+1}. \tag{3.19}$$

Observe that part (ii) of Theorems 3.3 and 3.4 follows from (3.17) and (3.18), respectively, together with some useful upper bounds of $c_{\varepsilon,N}$ and $c_{\varepsilon,D}$. However, to obtain (3.17) and (3.18), one must first establish part (iii) of Theorems 3.3 and 3.4, respectively, namely, the profiles of $u_{\varepsilon,N}$ and $u_{\varepsilon,D}$ (i.e., (3.11) and (3.16)). This turns out to rely heavily on some preliminary versions of (3.17) and (3.18). The proofs are indeed very involved, and we refer readers to the papers [NT2, NT3] and [NWe] for the full details. One interesting component in our proof of Theorem 3.4 is that the limit of δ_ε turns out to be a "viscosity" solution of the Hamilton–Jacobi equation

$$|\nabla\delta| = 1 \quad \text{in } \Omega,$$

which gives rise to the distance $d(P_{\varepsilon,D}, \partial\Omega)$. This also seems to indicate that although the function ϕ_ε (or V_ε) satisfies a simple linear elliptic equation while δ_ε satisfies a nonlinear one, the "correct" order of the error $(w - \mathcal{P}_{\Omega_\varepsilon} w)$ is far more important than the form of the equation it satisfies.

It turns out that the Neumann problem (3.1) also has a single-interior-peak spike-layer solution which is very close to the solution obtained in Theorem 3.4 (for the Dirichlet case (3.4)). We refer interested readers to [W2] or the next section.

Theorems 3.3 and 3.4 establish the existence of single-peak spike-layer solutions of (3.1) and (3.2), respectively. One natural question is, Are there other single-peak spike-layer solutions of (3.1) or (3.2)? And, if there are, where are the locations of their peaks?

For boundary spikes of the Neumann problem (3.1), Wei showed that if u_ε is a solution of (3.1) which has a single boundary peak P_ε, then, as $\varepsilon \to 0$, by passing to a subsequence if necessary, P_ε must tend to a critical point of the boundary mean curvature. On the other hand, for each *nondegenerate critical point* P_0 of the boundary mean curvature, one can always construct, for every $\varepsilon > 0$ small, a solution of (3.1) which has exactly one peak at $P_\varepsilon \in \partial\Omega$ such that $P_\varepsilon \to P_0$ as $\varepsilon \to 0$. (See [W1] for details.)

For interior spikes of the Neumann problem (3.1), it is established in [GPW] that (by passing to a subsequence if necessary) the single peak P_ε of a solution of (3.1) must tend to a *critical point* of the distance function $d(P, \partial\Omega)$. Conversely, again with additional hypotheses on Ω and a *nondegeneracy* condition on a critical point $P_0 \in \Omega$ of $d(P, \partial\Omega)$, one can construct, for every $\varepsilon > 0$ small, a solution of (3.1) which has exactly one peak at $P_\varepsilon \in \Omega$ such that $P_\varepsilon \to P_0$ as $\varepsilon \to 0$. (See [W2] for details.)

The counterparts of the above results for the Dirichlet case (3.2) are, however, not settled. Progress has been made in [W4].

3.1.2 Multipeak Spike-Layer Solutions in Elliptic Boundary-Value Problems

A vast amount of literature on (3.1) has been produced since the publication of [NT3] in 1993. Much progress has been made, and fascinating results concerning multipeak spike-layer solutions have been obtained. We will include only the most recent and complete results here. *Again, in this section we shall always assume that* $1 < p < \frac{n+2}{n-2}$ *in* (3.1).

An "ideal" result for multipeak spike-layer solutions to (3.1) would read as follows.

Conjecture. *For any given nonnegative integers k and ℓ,* (3.1) *always possesses a multipeak spike-layer solution with exactly k interior-peaks and ℓ boundary peaks, provided that ε is sufficiently small.*

The conjecture has almost been proved in this generality. In [GW2], this conjecture is established with some minor conditions imposed on the domain Ω. The main difficulty here comes from the fact the "error" in the boundary-peak case is *algebraic* (as shown in (3.11) and (3.17)), while the "error" in the interior-peak case is *transcendental* (as indicated in (3.16) or (3.18)). To overcome this, a delicate argument was devised in [GW2] to handle the gap in the error, but only under additional technical assumptions on Ω.

On the other hand, if we are to treat interior peaks and boundary peaks separately, definitive results are possible. For the case of interior peaks, the following result was obtained in [GW1].

Theorem 3.5. *Given any positive integer k, for every ε sufficiently small,* (3.1) *always possesses a multipeak spike-layer solution with exactly k interior peaks. Furthermore, as $\varepsilon \to 0$, the k peaks converge to a maximum point of the function*

$$\varphi(P^1,\ldots,P^k) = \min\left\{d(P^i,\partial\Omega), \frac{1}{2}|P^\ell - P^m| \,\middle|\, i,\ell,m = 1,\ldots,k\right\}, \tag{3.20}$$

where $P^1,\ldots,P^k \in \Omega$.

Intuitively speaking, a maximum point $(Q1,\ldots,Q^k)$ of the function φ in (3.20) corresponds to the centers of k disjoint balls of equal size packed in Ω (i.e., contained in Ω) with the largest possible diameter. Such a maximum point certainly exists, although it may not be unique.

The method of the proof is still variational; however, with the help of the Lyapunov–Schmidt reduction, the original "global" variational approach has now evolved into a powerful "localized" version. To illustrate the basic idea involved here, it seems best that we treat only the simplest case $k = 1$.

This "localized" energy method is semiconstructive. The strategy is simple: First, we construct an approximate solution in the sense of Lyapunov–Schmidt with its peak located at a prescribed point $P \in \Omega$. Then we perturb this point P and find a critical point of the corresponding "energy" of this approximate solution, which gives rise to an interior-peak spike-layer solution of (3.1).

To carry out this strategy, we let w be the solution of (3.10), as before, and, for any given point $P = (P_1,\ldots,P_n)$ in Ω, let $\mathcal{P}_{\varepsilon,P}w$ be the solution of

$$\begin{cases} \Delta v - v + w^p = 0 & \text{in } \Omega_{\varepsilon,P}, \\ \partial_\nu v = 0 & \text{on } \partial\Omega_{\varepsilon,P}, \end{cases}$$

where $\Omega_{\varepsilon,P} = \{z \in \mathbb{R}^n \,|\, x = P + \varepsilon z \in \Omega\}$. Now solve

$$u_{\varepsilon,P} = \mathcal{P}_{\varepsilon,P}w + \psi_{\varepsilon,P}, \tag{3.21}$$

with $\psi_{\varepsilon,P} \in K_{\varepsilon,P}^{\perp}$, where

$$K_{\varepsilon,P} = \operatorname{span}\left\{\frac{\partial \mathcal{P}_{\varepsilon,P}w}{\partial P_j} \,\middle|\, j = 1,\ldots,n\right\}, \tag{3.22}$$

and $\psi_{\varepsilon,P}$ is C^1 in P, $\Delta u_{\varepsilon,P} - u_{\varepsilon,P} + u_{\varepsilon,P}^p \in K_{\varepsilon,P}$, and

$$\|\psi_{\varepsilon,P}\|_{H^2(\Omega_{\varepsilon,P})} \le C\exp\left[-\frac{C}{\varepsilon}\, d(P,\partial\Omega)\right]. \tag{3.23}$$

Next, we define

$$\Phi_\varepsilon(P) = J_\varepsilon(u_{\varepsilon,P}) = J_\varepsilon(\mathcal{P}_{\varepsilon,P}w + \psi_{\varepsilon,P}). \tag{3.24}$$

It is not hard to see that a maximum point P_ε of Φ_ε, i.e.,

$$\Phi_\varepsilon(P_\varepsilon) = \max_{P\in\Omega} \Phi_\varepsilon(P), \tag{3.25}$$

gives a solution $u_\varepsilon = \mathcal{P}_{\varepsilon,P_\varepsilon} w + \psi_{\varepsilon,P_\varepsilon}$ of (3.1) because

$$\Delta u_\varepsilon - u_\varepsilon + u_\varepsilon^p = \sum_{j=1}^{n} \alpha_j \frac{\partial \mathcal{P}_{\varepsilon,P_\varepsilon} w}{\partial P_j} \tag{3.26}$$

for some $\alpha_1, \ldots, \alpha_n \in \mathbb{R}$. From (3.25) and (3.26) it follows that

$$\begin{aligned} 0 &= \frac{\Phi_\varepsilon(P)}{\partial P_k}\bigg|_{P=P_\varepsilon} = J_\varepsilon'(u_\varepsilon)\frac{\partial u_\varepsilon}{\partial P_k} \\ &= \sum_{j=1}^{n} \alpha_j \int_{\Omega_{\varepsilon,P_\varepsilon}} \frac{\partial \mathcal{P}_{\varepsilon,P_\varepsilon} w}{\partial P_j} \frac{\partial(\mathcal{P}_{\varepsilon,P_\varepsilon} w + \psi_{\varepsilon,P_\varepsilon})}{\partial P_k} = \sum_{j=1}^{n} a_{kj}\alpha_j, \end{aligned} \tag{3.27}$$

where the matrix (a_{kj}) is defined by the last equality. Due to the cancellation property of the integral

$$\int_{\Omega_{\varepsilon,P_\varepsilon}} \frac{\partial \mathcal{P}_{\varepsilon,P_\varepsilon} w}{\partial P_j} \frac{\partial \mathcal{P}_{\varepsilon,P_\varepsilon} w}{\partial P_k} = 0, \quad k \neq j, \tag{3.28}$$

and (3.23), we see that a_{kk} is much larger than a_{kj}, $k \neq j$. Consequently, $\det(a_{kj}) \neq 0$. Therefore, $\alpha_j = 0, j = 1, \ldots, n$, and the proof is complete. This method generalizes to arbitrary $k > 1$.

For the case of boundary peaks, the situation becomes more interesting. The following result is obtained in [GWW].

Theorem 3.6. *Given any positive integer k, (3.1) always possesses a multipeak spike-layer solution with exactly k boundary peaks, provided that ε is sufficiently small. Furthermore, as $\varepsilon \to 0$, the k peaks $Q_1^\varepsilon, \ldots, Q_k^\varepsilon$ have the following property:*

$$H(Q_j^\varepsilon) \to \min_{P \in \partial\Omega} H(P),$$

where H denotes the mean curvature of $\partial\Omega$.

Comparing the above result to Theorem 3.3, we see a very interesting difference in the location of the peaks: Theorem 3.3 guarantees the existence of a single boundary peak near a maximum of the boundary mean curvature, while Theorem 3.6 implies the existence of an arbitrary number of boundary peaks near a minimum of the boundary mean curvature. Whether (3.1) has a spike-layer solution with exactly k boundary peaks near a maximum of the boundary mean curvature, for a prescribed positive integer k, remains an interesting open question.

The proof uses basically the same approach as that of Theorem 3.5. The detailed computations are, of course, quite different. We refer interested readers to [GWW] for details.

In [GW1], [GPW], it is also proved that if $P_\varepsilon^1, \ldots, P_\varepsilon^k$ in Ω are the locations of the k (interior) peaks of a solution to (3.1), then, by passing to a subsequence if necessary, $P_\varepsilon^1, \ldots, P_\varepsilon^k$ must tend to a critical point of φ in (3.20).

We see that we have acquired a fairly good understanding of spike-layer solutions of (3.1), although there are still major questions left open. On the other hand, our knowledge of solutions of (3.1) with multidimensional concentration sets is very limited at this time. In the next section we shall report some progress in that direction.

To conclude this section we remark that the existence of multipeak spike-layer solutions for the Dirichlet problem (3.2) is, in general, not possible. For instance, when Ω is a ball, [GNN1] implied that solutions of (3.2) must be radially symmetric and thus (3.2) can have only single-peak solutions. This result is extended to strictly convex domains by Wei [W2]. Therefore, the existence of multipeak spike-layer solutions for the Dirichlet problem (3.2) is drastically different from its counterpart of the Neumann case (3.1) and generally depends on the geometry of Ω.

3.1.3 Solutions with Multidimensional Concentration Sets

A spike-layer solution (discussed earlier) has the property that its "energy" or "mass" concentrates near isolated points (i.e., its peaks) in $\overline{\Omega}$, which are zero dimensional. Therefore we view *a spike-layer solution as a solution with a zero-dimensional concentration set.* Similarly, solutions which are small everywhere except near a curve or curves are regarded as *solutions with one-dimensional concentration sets.* Generalizing in this manner, we can define *solutions with k-dimensional concentration sets.*

The following conjecture has been around for quite some time.

Conjecture. *Given any integer $0 \leq k \leq n-1$, there exists $p_k \in (1,\infty]$ such that for $1 < p < p_k$, (3.1) possesses a solution with k-dimensional concentration set, provided that ε is sufficiently small.* (See, for instance, [N2].)

Progress in this direction has been made only very recently. In [MM1, MM2] the above conjecture was established for a sequence of $\varepsilon \to 0$ in the case $k = n-1$ with the boundary $\partial\Omega$ (or, part of $\partial\Omega$) being the concentration set.

Theorem 3.7. *Let $\Omega \subseteq \mathbb{R}^n$ be a bounded smooth domain, and let $p > 1$. Then, for any component Γ of $\partial\Omega$, there exists a sequence $\varepsilon_m \to 0$ such that (3.1) possesses a solution u_{ε_m} for $\varepsilon = \varepsilon_m$ and u_{ε_m} concentrates at Γ; i.e., $u_{\varepsilon_m} \to 0$ away from Γ and $u_{\varepsilon_m}(\varepsilon_m(x - x_0)) \to w(x \cdot \nu_0)$ near $x_0 \in \Gamma$, where ν_0 is the unit inner normal to Γ at x_0 and w is the solution of (3.10) with $n = 1$.*

Note that in Theorem 3.7, no upper bound is imposed on p. The proof is very interesting, though technical, and we shall give only a very brief outline. The first crucial step is to construct a "good" approximate solution $\widetilde{u}_\varepsilon$. Then, a detailed analysis of the second differential $J''_{\varepsilon,N}(\widetilde{u}_\varepsilon)$, where $J_{\varepsilon,N}$ is defined in (3.6), is essential for using the contraction mapping argument to obtain the solution u_ε near $\widetilde{u}_\varepsilon$. In the second step it is observed that the Morse index of $\widetilde{u}_\varepsilon$ tends to ∞ as $\varepsilon \to 0$, and thus the invertibility of $J''_{\varepsilon,N}(\widetilde{u}_\varepsilon)$ is guaranteed only along a sequence $\varepsilon_m \to 0$. Similar to the proofs earlier, the construction of approximate solutions here is crucial, and is delicate and interesting. Roughly speaking, it is natural to use the one-dimensional solution of (3.10) as a candidate for approximate solution. This turns out to be inadequate. In [MM1], (3.10) was replaced by

$$\begin{cases} w'' - w + w^p = -\lambda w' & \text{in } (0,\infty), \\ w'(0) = w(\infty) = 0 & \text{and } w > 0, \end{cases} \tag{3.29}$$

where λ is related to the mean curvature of the boundary of $\Omega_\varepsilon = \frac{1}{\varepsilon}\Omega$ (which tends to 0 as $\varepsilon \to 0$).

A natural question would be whether (3.1) admits any solutions which concentrate on "interior" curves, and if there are such solutions, how the locations of these "interior" curves are determined. Even at the formal level, these are difficult questions. To this date, progress has been made only in very special cases.

In [AMN3], the following result was established.

Theorem 3.8. *Let Ω be the unit ball B_1 in $\mathbb{R}^n$. Then, for every $p > 1$ and ε small,* (3.1) *possesses a radial solution u_ε concentrating at $|x| = r_\varepsilon$ for which $1 - r_\varepsilon \sim \varepsilon|\log\varepsilon|$.*

(Here we use the notation "$f \sim g$" to denote that as $\varepsilon \to 0$, the quotient f/g is bounded from above and below by two positive constants.) Note that again we do not impose any upper bound on p here.

We remark that the solution guaranteed by Theorem 3.8 is different from the one in Theorem 3.7, as the maximum of the solution in Theorem 3.7 takes place *on the boundary*, while the maximum of the solution in Theorem 3.8 lies *in the interior of* Ω. In fact, it is possible to construct a solution of (3.1) which concentrates on *a cluster of spheres*.

Theorem 3.9 (see [MNW]). *Let Ω be the unit ball B_1 in $\mathbb{R}^n$ and N be a given positive integer. Then, for every $p > 1$ and ε small,* (3.1) *possesses a radial solution u_ε concentrating on N spheres $|x| = r_{\varepsilon,j}$, $j = 1,\ldots,N$, where $1 - r_{\varepsilon,1} \sim \varepsilon|\log\varepsilon|$ and $r_{\varepsilon,j} - r_{\varepsilon,j+1} \sim \varepsilon|\log\varepsilon|$ for $j = 1,\ldots,N-1$.*

Since the basic ideas involved in Theorems 3.8 and 3.9 are similar, we shall confine our discussions in the rest of this section to Theorem 3.8 for the sake of simplicity.

It is obvious that (3.2)—the Dirichlet counterpart of (3.1)—does not admit any solutions other than a single-peak solution in case Ω is a ball, as is guaranteed by [GNN1]. Nevertheless, Theorem 3.8 gives us a good opportunity to compare Dirichlet and Neumann boundary conditions. To illustrate the ideas involved, it seems best to discuss a slightly more general equation,

$$\varepsilon^2\Delta u - V(|x|)u + u^p = 0 \text{ and } u > 0 \text{ in } B_1, \tag{3.30}$$

under the boundary conditions

$$\partial_\nu u = 0 \quad \text{on } \partial\Omega \tag{3.31}$$

or

$$u = 0 \quad \text{on } \partial\Omega, \tag{3.32}$$

where V is a radial potential bounded by two positive constants. In fact, the relevant quantity turns out to be

$$M(r) = r^{n-1}V^\theta(r),\ \theta = \frac{p+1}{p-1} - \frac{1}{2}, \tag{3.33}$$

and Theorem 3.8 is a special case of the following result [AMN3].

Theorem 3.10.

(i) *If $M'(1) > 0$, then for every $p > 1$ and ε small, the problem* (3.30), (3.31) *possesses a solution u_ε concentrating at $|x| = r_\varepsilon$, where $1 - r_\varepsilon \sim \varepsilon|\log\varepsilon|$.*

(ii) *If $M'(1) < 0$, then for every $p > 1$ and ε small, the problem* (3.30), (3.32) *possesses a solution u_ε concentrating at $|x| = r_\varepsilon$, where $1 - r_\varepsilon \sim \varepsilon|\log\varepsilon|$.*

Remark. In addition to the solution in Theorem 3.10, the problems (3.30), (3.31) and (3.30), (3.32) also have solutions concentrating near $|x| = \bar{r}$, where (and if) $\bar{r}$ is a local extreme point of M. This particular solution also exists for the nonlinear Schrödinger equations in $\mathbb{R}^n$. (See [AMN2].)

The proof relies upon a finite-dimensional Lyapunov–Schmidt reduction and a "localized" energy method. Again, the first crucial step, for both (i) and (ii), is to find a good approximate solution $z^\varepsilon_{\rho,N}$ for (i) and $z^\varepsilon_{\rho,D}$ for (ii), where ρ is a parameter between 0 and $1/\varepsilon$, denoting the radius of concentration and will be determined eventually. Observe that a solution to the problem (3.30) and (3.31) is a critical point of the (rescaled) functional on $H_{1,r}(B_{1/\varepsilon})$,

$$\widetilde{J}_{\varepsilon,N}(u) = \frac{1}{2}\int_{B_{1/\varepsilon}} \left(|\nabla u|^2 + V(\varepsilon|x|)u^2\right) - \frac{1}{p+1}\int_{B_{1/\varepsilon}} u_+^{p+1},$$

where $H_{1,r}$ is the space of all radial H_1 functions on $B_{1/\varepsilon}$. The general abstract procedure establishing $\widetilde{J}'_{\varepsilon,N}(z^\varepsilon_{\rho,N} + w) = 0$ is equivalent to

(a) finding $w = w^\varepsilon_{\rho,N} \in (T_z Z^\varepsilon_N)^\perp$ such that $P\widetilde{J}'_{\varepsilon,N}(z^\varepsilon_{\rho,N} + w) = 0$, and
(b) finding a stationary point of

$$\Psi_{\varepsilon,N}(\rho) = \widetilde{J}_{\varepsilon,N}(z^\varepsilon_{\rho,N} + w^\varepsilon_{\rho,N}),$$

where P denotes the orthogonal projection from $H_{1,r}$ onto $(T_z Z^\varepsilon_N)^\perp$, $T_z Z^\varepsilon_N$ is the tangent space of Z^ε_N at z, and Z^ε_N is the family of the approximate solutions $z^\varepsilon_{\rho,N}$. To find a nontrivial critical value of $\Psi_{\varepsilon,N}$, we have the expansion

$$\varepsilon^{n-1}\Psi_{\varepsilon,N}(\rho) = M(\varepsilon\rho)\left(\alpha - \beta e^{-2\lambda(\frac{1}{\varepsilon}-\rho)}\right) + \text{higher-order terms}, \tag{3.34}$$

where α, β are two positive constants and $\lambda^2 = V(\varepsilon\rho)$. Now, differentiating (3.34) with respect to ρ and setting the leading term to 0, we obtain

$$\varepsilon M'(\varepsilon\rho)\left(\alpha - \beta e^{-2\lambda(\frac{1}{\varepsilon}-\rho)}\right) + 2\beta M(\varepsilon\rho)\left[\varepsilon\lambda'(\varepsilon\rho)\left(\frac{1}{\varepsilon} - \rho\right) - \lambda(\varepsilon\rho)\right]e^{-2\lambda(\frac{1}{\varepsilon}-\rho)} = 0.$$

If $\frac{1}{\varepsilon} - \rho \sim |\log\varepsilon|$, then as $\varepsilon \to 0$

$$e^{-2\lambda(\frac{1}{\varepsilon}-\rho)} \to 0,$$

and therefore

$$\varepsilon\alpha M'(\varepsilon\rho) \sim 2\beta M(\varepsilon\rho)\lambda(\varepsilon\rho)e^{-2\lambda(\frac{1}{\varepsilon}-\rho)},$$

which can be achieved if $M'(1) > 0$ (since $\varepsilon\rho \to 1$ and $\frac{1}{\varepsilon} - \rho \sim |\log\varepsilon|$).

For the Dirichlet case (3.30) and (3.32) we define similarly the functionals $\widetilde{J}_{\varepsilon,D}$ and $\Psi_{\varepsilon,D}$ and the expansion corresponding to (3.34) now reads as follows:

$$\varepsilon^{n-1}\Psi_{\varepsilon,D}(\rho) = M(\varepsilon\rho)\left(\alpha + \beta e^{-2\lambda(\frac{1}{\varepsilon}-\rho)}\right) + \text{higher-order terms}. \tag{3.35}$$

Comparing (3.35) to (3.34), we see that only the sign for the term $\beta e^{-2\lambda(\frac{1}{\varepsilon}-\rho)}$ is different, which reflects the different, or opposite, effects of Dirichlet and Neumann boundary conditions. Heuristically, when $V \equiv 1$, the first term in (3.34) and (3.35) is due to the volume energy which always has a tendency to "shrink" (in order to minimize), while the second term $\pm\beta e^{-2\lambda(\frac{1}{\varepsilon}-\rho)}$ in (3.35) and (3.34), respectively, indicates that in the Dirichlet case the boundary "pushes" the mass of the solution away from the boundary (therefore only single-peak solutions are possible), but in the Neumann case the boundary "pulls" the mass of the solution and thereby reaches a balance at $\varepsilon\rho = r_\varepsilon \sim 1 - \varepsilon|\log\varepsilon|$ creating an extra solution.

We remark that the method described above also applies to the annulus case and yields the following interesting results for $V \equiv 1$, which illustrate the opposite effects between Dirichlet and Neumann boundary conditions most vividly.

Theorem 3.11 (see [AMN3]).

(i) *For every $p > 1$ and ε small, the Neumann problem* (3.1) *with $\Omega = \{x \in \mathbb{R}^n \,|\, 0 < a < |x| < b\}$ possesses a solution concentrating at $|x| = r_\varepsilon$, where $b - r_\varepsilon \sim \varepsilon|\log\varepsilon|$, near the outer boundary $|x| = b$.*

(ii) *For every $p > 1$ and ε small, the Dirichlet problem* (3.2) *with $\Omega = \{x \in \mathbb{R}^n \,|\, 0 < a < |x| < b\}$ possesses a solution concentrating at $|x| = r_\varepsilon$, where $r_\varepsilon - a \sim \varepsilon|\log\varepsilon|$, near the inner boundary $|x| = a$.*

Observe that from the "moving plane" method [GNN1] it follows easily that the Dirichlet problem (3.2) does not have a solution concentrating on a sphere near the outer boundary $|x| = b$.

In conclusion, we mention that the method of Theorem 3.10 can be extended to produce solutions with k-dimensional concentration sets, but again, some symmetry assumptions are needed. Other interesting progress in this direction includes a one-dimensional concentration set in the interior of a two-dimensional domain due to [WY]. The conjecture stated at the beginning of this section remains largely a major open problem.

3.1.4 Remarks

In Section 3.1, we have considered the various concentration phenomena for essentially just one equation, namely,

$$\varepsilon^2\Delta u - u + u^p = 0 \tag{3.36}$$

in a bounded domain Ω under either Dirichlet or Neumann boundary conditions in (3.2) or (3.1), respectively. However, since (3.36) is quite basic, similar phenomena could be expected for a more general class of equations. We again mention that, as ε becomes large, (3.1) will eventually lose all of its solutions except the trivial one $u_\varepsilon \equiv 1$ [LNT].

The methods involved in handling (3.36) are basically variational, more precisely, via the Mountain-Pass lemma. However, the Mountain-Pass approach we have used here is due to Ding and Ni [DN] in 1983, which deviates from the original one due to Ambrosetti and Rabinowitz [AR] and is less general but more constructive. As a result, it is proved in [N1] that this approach yields the same critical value as the constrained minimization approach due to Nehari [Ne] in 1960. In studying multipeak solutions and other concentration sets, this approach has been modified; namely, it is also coupled with the Lyapunov–Schmidt finite-dimensional reduction and becomes "local" in nature. This "localized energy method" is a major achievement, and is due to Gui and Wei [GW1].

It is interesting to note that the concentration sets of solutions to (3.1) we have discussed so far have dimensions ranging from 0 (i.e., peaks) to $n-1$ (spheres in $\mathbb{R}^n$). A natural question arises: Does (3.1) possess solutions with n-dimensional concentration sets? In general, this question remains open, although the answer is probably negative.

Solutions with n-dimensional concentration sets (often referred to as internal transition layers) appear in phase transitions. This problem has been studied extensively in the past 30 years by many authors, including Alikakos, Bates, Xinfu Chen, Fife, Fusco, Hale, Mimura, Sakamoto, and others. We refer interested readers to the monograph [F] for further details.

A "nonautonomous" version of (3.1) was considered by Ren [R],

$$\begin{cases} \varepsilon^2\Delta u - u + K(x)u^p = 0 & \text{in } \Omega, \\ u > 0 & \text{in } \Omega, \\ \partial_\nu u = 0 & \text{on } \partial\Omega, \end{cases} \tag{3.37}$$

where $1 < p < \frac{n+2}{n-2}$ and $K(x) > 0$. Results similar to those in Theorem 3.3 were obtained, and for ε small the location of the peak is determined by an interesting interaction between the maximum points of K on $\partial\Omega$ and in Ω.

3.2 Concentrations of Solutions: Systems

In this section, we shall return to the activator-inhibitor system (3.3) discussed in Section 3.1. Our first goal is to construct stationary solutions of (3.3) for large d_2 given the knowledge of the single-peak spike-layer solutions of (3.1) in Section 3.1. It turns out that this is accomplished only under additional assumptions.

In Subsection 3.2.2, we include two more systems: the CIMA reaction and the Gray–Scott model. Both present extremely rich and interesting phenomena in pattern formations, and many mathematical questions remain open.

If spatial heterogeneity is considered, especially when interacting with diffusion, very interesting phenomena often occur. We shall discuss those in Chapters 4 and 5. Here in this section we confine our attention to "autonomous" equations and systems.

3.2.1 The Gierer–Meinhardt System

In the one-dimensional case, much is known about the system (3.3) due to the work of Takagi [Ta]. We shall therefore focus on the case $n \geq 2$ in this section. The existence question for nontrivial stationary solutions to the activator-inhibitor system (3.3) under the condition (3.4) for general domain Ω remains open. Progress has been made, and there are two approaches to this problem. The first one is via the shadow system. The best result in this direction so far seems to be [NT4], in which the domain Ω is assumed to be axially symmetric and *multipeak spike-layer steady states* are constructed. Here we are going to give a brief description of this approach.

The steady states of (3.3) satisfy the following elliptic system:

$$\begin{cases} d_1\Delta U - U + \dfrac{U^p}{V^q} = 0 & \text{in } \Omega, \\ d_2\Delta V - V + \dfrac{U^r}{V^s} = 0 & \text{in } \Omega, \\ U > 0 \text{ and } V > 0 & \text{in } \Omega, \\ \partial_\nu U = 0 = \partial_\nu V & \text{on } \partial\Omega, \end{cases} \tag{3.38}$$

where p, q, r are positive, $s \geq 0$, and

$$0 < \frac{p-1}{q} < \frac{r}{s+1}. \tag{3.39}$$

Similarly, the elliptic "shadow" system is

$$\begin{cases} d_1\Delta U - U + \dfrac{U^p}{\xi^q} = 0 & \text{in } \Omega, \\ \int_\Omega U^r = |\Omega|\xi^{s+1}, & \\ \partial_\nu U = 0 & \text{on } \partial\Omega. \end{cases} \tag{3.40}$$

If we set

$$U(x) = \xi^{\frac{q}{p-1}} u(x),$$

then (3.40) is equivalent to

$$\begin{cases} \varepsilon^2\Delta u - u + u^p = 0 & \text{in } \Omega, \\ \int_\Omega u^r = |\Omega|\xi^{-\alpha}, & \\ \partial_\nu u = 0 & \text{on } \partial\Omega, \end{cases} \tag{3.41}$$

where $d_1 = \varepsilon^2$ and

$$\alpha = \frac{qr}{p-1} - (s+1) > 0. \tag{3.42}$$

Now, suppose that the x_n-axis is the axis of symmetry for Ω and that $P_1, \ldots, P_{2N}$ are the points at which $\partial\Omega$ intersects the x_n-axis. The following result is proved in [NT4].

Theorem 3.12. *Under the assumption* (3.42), *given any m distinct points $P_{j_1}, \ldots, P_{j_m} \in \{P_1, P_2, \ldots, P_{2N}\}$, there are two constants D_1 and D_2 such that for every $0 < d_1 < D_1$ and $d_2 > D_2$ the system* (3.38) *has a spike-layer solution with exactly m peaks at $P_{j_1}, \ldots, P_{j_m}$.*

To illustrate our approach, we shall treat only the case $m = 1$, as the general case $m > 1$ requires no new ideas or techniques.

First, we introduce a diffeomorphism in order to flatten a boundary portion around $P \in \{P_1, \ldots, P_{2N}\}$ as follows. Assuming P is the origin, we see that there is a smooth function $\psi \in C^\infty([-\delta,\delta])$ with $\psi(0) = \psi'(0) = 0$ such that, near P, $\partial\Omega$ is represented by $\{(x', \psi(|x'|) \mid |x'| < \delta\}$. Setting

$$\Phi_j(y) = \begin{cases} y_j - y_n \psi'(|y'|)\frac{y_j}{|y'|}, & j = 1, \ldots, n-1, \\ y_n + \psi(|y'|), & j = n, \end{cases} \tag{3.43}$$

we see that $x = \Phi(y) = (\Phi_1(y), \ldots, \Phi_n(y))$ is a diffeomorphism from an open set containing the closed ball $\overline{B}_{3_\kappa}$, where $\kappa > 0$ is small, onto a neighborhood of P with $D\Phi(0) = 1$, the identity map. Observe that $x = \Phi(y)$ maps the hyperplane $\{y_n = 0\}$ into $\partial\Omega$. We write $\Psi = \Phi^{-1}$.

Now we can write

$$u(x) = \chi(\kappa^{-1}|\Psi(x)|) \cdot w(\varepsilon^{-1}\Psi(x)) + \varepsilon\phi \equiv \widetilde{u}_\varepsilon + \varepsilon\phi, \tag{3.44}$$

where w is the solution of (3.10) and $\chi \in C_0^\infty(\mathbb{R})$ is a cut-off function such that (i) $0 \le \chi(s) \le 1$, (ii) $\chi(s) = 1$ if $|s| \le 1$, and (iii) $\chi(s) = 0$ if $|s| \ge 2$. Note that $\widetilde{u}$ is an approximate solution of the equation in (3.41) and the equation in (3.41) now takes the following form:

$$L_\varepsilon\phi + g_\varepsilon + M_\varepsilon[\phi] = 0,$$

where

$$L_\varepsilon = \varepsilon^2\Delta - 1 + p\widetilde{u}_\varepsilon^{p-1},$$

$$g_\varepsilon = \frac{1}{\varepsilon}[\varepsilon^2\Delta\widetilde{u}_\varepsilon - \widetilde{u}_\varepsilon + \widetilde{u}_\varepsilon^p],$$

$$M_\varepsilon[\phi] = \frac{1}{\varepsilon}[(\widetilde{u}_\varepsilon + \varepsilon\phi)^p - \widetilde{u}_\varepsilon^p - \varepsilon p\widetilde{u}_\varepsilon^{p-1}\phi].$$

It turns out that $M_\varepsilon[\phi]$ is small and g_ε is bounded. While L_ε is not invertible in general, it actually has a bounded inverse when restricted to *axially symmetric* functions. This enables us to solve the equation in (3.41) with a solution of the form (3.44). Then we simply define

$$\xi_\infty = |\Omega|^{\frac{1}{\alpha}} \left(\int_\Omega u^r \right)^{-\frac{1}{\alpha}}$$

and $U_\infty(x) = \xi_\infty^{q/(p-1)} u(x)$, and we obtain a solution (U, ξ) of the shadow system (3.40).

The original system (3.38) with d_2 large turns out to be a *regular perturbation* of the shadow system (3.40). If we write $\delta = d_2^{-1}$ and define the operator

$$\mathcal{P}u = u - \frac{1}{|\Omega|}\int_\Omega u,$$

then we can convert solving the system (3.38) to finding a zero (U, ξ, ϕ) of the map $\mathcal{F} = (\mathcal{F}_1, \mathcal{F}_2, \mathcal{F}_3)$ for $\delta > 0$ but small, where

$$\mathcal{F}_1(U, \xi, \phi; \delta) = \varepsilon^2\Delta U - U + \frac{U^p}{(\xi + \phi)^q},$$

$$\mathcal{F}_2(U,\xi,\phi;\delta) = \int_\Omega \left[-(\xi+\phi) + \frac{U^r}{(\xi+\phi)^s} \right],$$

$$\mathcal{F}_3(U,\xi,\phi;\delta) = \Delta\phi + \delta\left[-\phi + \mathcal{P}\left(\frac{U^r}{(\xi+\phi)^s} \right) \right],$$

near $(U_\infty, \xi_\infty, 0)$ (the solution for (3.40), corresponding to the case $\delta = 0$). Notice that we have decomposed the second equation in (3.38) into two equations so that the linearization of the map $\mathcal{F}$ at $(U_\infty, \xi_\infty, 0; 0)$ is invertible in suitable function spaces and thereby (3.38) can be solved by the Implicit Function theorem.

The second approach is due to Wei and his collaborators. In this approach, d_2 need not be very large, although there are other restrictions; in particular, this approach works only for planar domains, i.e., $n = 2$. To illustrate the basic idea involved here, we take the special case $s = 0$ in (3.38). The first step here is to solve the second equation in (3.38):

$$\begin{cases} d_2\Delta V - V + U^r = 0 & \text{in } \Omega, \\ \partial_\nu V = 0 & \text{on } \partial\Omega. \end{cases} \tag{3.45}$$

Then, writing $V = T[U^r]$ and substituting into the first equation in (3.38), we have

$$\begin{cases} d_1\Delta U - U + \dfrac{U^p}{(T[U^r])^q} = 0 & \text{in } \Omega, \\ \partial_\nu U = 0 & \text{on } \partial\Omega. \end{cases}$$

It is observed that, under suitable scalings, (3.45) will have a solution close to a large constant, namely,

$$V \sim \xi_\varepsilon \left(1 + O(|\log\varepsilon|^{-1}) \right),$$

where $d_1 = \varepsilon^2$ is small and $\xi_\varepsilon \to \infty$ as $\varepsilon \to 0$. In this approach, the asymptotic behavior of the Green's function

$$\begin{cases} \Delta G - G + \delta_P = 0 & \text{in } \Omega, \\ \partial_\nu G = 0 & \text{on } \partial\Omega, \end{cases}$$

where δ_P denotes the Dirac δ-function at the point P, is essential, which limits this approach to $n = 2$ only. (See [WW1, WW2].)

3.2.2 Other Systems

The 1952 paper of Turing [T], in which the novel notion of "diffusion-driven instability" was first proposed in an attempt to model the regeneration phenomenon of *hydra*, is one of the most important papers in theoretical biology in the last century. However, the two chemicals, activator and inhibitor in Turing's theory, are yet to be identified in *hydra*.

The first experimental evidence of Turing pattern was observed in 1990, nearly 40 years after Turing's prediction, by the Bordeaux group in France on the chlorite-iodide-malonic acid-starch (CIMA) reaction in an open unstirred gel reactor [CDBD]. In their scheme, the two sides of the gel strip loaded with starch indicator are, respectively, in contact with solutions of chlorite (CLO_2^-) and iodide (I^-) ions on one side, and malonic acid (MA) on the other side, which are fed through two continuous-flow stirred tank reactors.

These reactants diffuse into the gel, encountering each other at significant concentrations in a region near the middle of the gel, where the Turing patterns of lines of periodic spots can be observed. This observation represents a significant breakthrough for one of the most fundamental ideas in morphogenesis and biological pattern formation.

The Brandeis group later found that, after a relatively brief initial period, it is really the simpler chlorine dioxide CLO_2-I_2-MA (CDIMA) reaction that governs the formation of the patterns [LE1, 2]. The CDIMA reaction can be described in a five-variable model that consists of three component processes. However, observing that three of the five concentrations remain nearly constants in the reaction, Lengyel and Epstein [LE1, LE2] simplified the model to a 2×2 system: Let $U = U(x,t)$ and $V = V(x,t)$ denote the chemical concentrations (rescaled) of iodide (I^-) and chlorite (CLO_2^-), respectively, at time t and $x \in \Omega$, where Ω is a smooth, bounded domain in $\mathbb{R}^n$. Then the Lengyel and Epstein model takes the form

$$\begin{cases} U_t = \Delta U + a - U - \dfrac{4UV}{1+U^2} & \text{in } \Omega \times (0,T), \\ V_t = \sigma \left[c\Delta V + b\left(U - \dfrac{UV}{1+U^2} \right) \right] & \text{in } \Omega \times (0,T), \\ \partial_\nu U = \partial_\nu V = 0 & \text{on } \partial\Omega \times (0,T), \end{cases} \tag{3.46}$$

where a and b are parameters related to the feed concentrations, c is the ratio of the diffusion coefficients, and $\sigma > 1$ is a rescaling parameter depending on the concentration of the starch, enlarging the effective diffusion ratio to σc. All constants a, b, c, and σ are assumed to be positive.

It was established in [NTa] that all solutions of (3.46) must eventually enter the region

$$R_a = (0,a) \times (0, 1 + a^2)$$

for t large, regardless of the initial values $U(x,0)$, $V(x,0)$. Furthermore, the existence and nonexistence of steady states of (3.46) are also investigated in [NTa]. Results there show that, roughly speaking, *if any one of the three quantities*

(i) *the parameter a (related to the feed concentrations),*

(ii) *the size of the reactor* Ω *(reflected by its first eigenvalue),*

(iii) *the "effective" diffusion rate* $d = c/b$

is not large enough, then the system (3.46) *has no nonconstant steady states.*

On the other hand, it was also established in [NTa] that *if a lies in a suitable range, then* (3.46) *possesses nonconstant steady states for large d*. The proof of the existence uses a degree-theoretical approach combined with the a priori bounds. However, such an approach does not provide much information about the shape of the solution. In the case $n = 1$, a better description for the structure of the set of nonconstant steady states to (3.46) is given in [JNT]; namely, a global bifurcation theorem which gives the existence of nonconstant steady states to (3.46) for all d suitable large (under a rather natural condition) is obtained. Moreover, the corresponding shadow system (as $d \to \infty$) is also solved in [JNT].

There are various experimental and numerical studies on the system (3.46); see, e.g., [CK], [JS] and the references therein. However, the qualitative properties of solutions to (3.46) remain largely open.

Another system supporting many interesting spatio-temporal patterns is the Gray–Scott model [GS]. It models an irreversible autocatalytic chemical reaction involving two reactants in a gel reactor, where the reactor is maintained in contact with a reservoir of one of the two chemicals in the reaction. In dimensionless units it can be written as

$$\begin{cases} U_t = D_U \Delta U + F(1-U) - UV^2 & \text{in } \Omega \times (0,T), \\ V_t = D_V \Delta V - (F+k)V + UV^2 & \text{in } \Omega \times (0,T), \\ \partial_\nu U = \partial_\nu V = 0 & \text{on } \partial\Omega \times (0,T), \end{cases} \tag{3.47}$$

where $U = U(x,t)$ and $V = V(x,t)$ represent the concentrations of the two chemicals at point $x \in \Omega \subset \mathbb{R}^n, n \leq 3$, and at time $t > 0$, respectively, and D_U, D_V are the diffusion coefficients of U and V, respectively. F denotes the rate at which U is fed from the reservoir into the reactor, and k is a reaction-time constant.

For various ranges of these parameters, (3.47) is expected to admit a rich solution structure involving pulses or spots, rings, stripes, self-replication spots, and spatio-temporal chaos. See [Pe] and [LMPS] for numerical simulations and experimental observations.

In one-dimensional case $n = 1$, a stationary one-pulse solution in the *entire real line* (i.e., $\Omega = \mathbb{R}^1$ and no boundary condition imposed in (3.47)) is studied in [DGK]. In case $n = 2$, "spotty" solutions are investigated in [W3] and [WW3]. Many other patterns here remain to be established with mathematical rigor.

3.3 Symmetry and Related Properties of Solutions

Symmetry of solutions to elliptic problems has been an important subject in mathematics as well as in mathematical physics. It has a long history, and the early works dealt mainly with symmetric rearrangements [PS]. Systematic studies of symmetry properties in modern era only seem to have started with the papers [GNN1, GNN2] around 1980.

The basic question here is, *If the elliptic equation, the underlying domain, and the boundary conditions all possess certain symmetries, then do all positive solutions have the same symmetries? Or, what kind of symmetries would positive solutions possess?*

Spherical symmetry, or radial symmetry, being among the most fundamental symmetries, will be the center of this section.

We shall start with the simplest semilinear elliptic equation,

$$\Delta u + f(u) = 0 \quad \text{in } \Omega, \tag{3.48}$$

where Ω is a ball, an annulus, or the entire Euclidean space $\mathbb{R}^n$, with either the homogeneous Dirichlet boundary condition

$$u = 0 \quad \text{on } \partial\Omega \tag{3.49}$$

or the homogeneous Neumann boundary condition

$$\partial_\nu u = 0 \quad \text{on } \partial\Omega. \tag{3.50}$$

It will become abundantly clear that the Dirichlet condition (3.49) is far more "coercive" or "imposing" than the Neumann boundary condition (3.50). More precisely, the Dirichlet condition (3.49) will "force" all positive solutions of (3.48) to posses the same radial symmetry in the case when Ω is a ball, while the Neumann condition (3.50) allows many other possibilities. These are discussed in Subsection 3.3.1.

It is interesting to note that although annuli also have the same radial symmetry, positive solutions of (3.48) in an annulus even under the Dirichlet boundary condition (3.49) are generally *not* radially symmetric. In fact, the "least-energy" solutions of (3.48) and (3.49) discussed in Section 3.1 are often *not* radially symmetric. This indicates that, for an annulus, "natural" solutions (or "ground states") to (3.48) and (3.49) often do *not* inherit radial symmetry from the domain in order to stay "more stable," i.e., in order to reduce its "energy." A more detailed explanation and other relevant information are included in Subsection 3.3.2.

For the case $\Omega = \mathbb{R}^n$, we usually consider only positive solutions which tend to 0 at ∞. This, of course, seems weaker than the Dirichlet boundary condition (3.49) for the ball case. Indeed, symmetry properties in this case are generally more difficult to establish. Subsection 3.3.3 is devoted to this case.

The main method used in this section is the "moving plane" method, which is based on the Maximum Principle and a device due to A. D. Alexandroff in 1956 in his proof of the result that *all topological spheres in* $\mathbb{R}^n$ *with constant mean curvatures must actually be spheres*. This method establishes radial symmetry by first proving mirror symmetry, or reflection symmetry, in an arbitrary direction. In the process it also yields the monotonicity of solutions as a by-product. Since the publication of [GNN1, GNN2], there have been many variants of this "moving plane" method developed by various authors to handle different types of domains, such as half-spaces, and/or to establish related properties of solutions, such as monotonicity properties. One such example is the De Giorgi conjecture, which will be discussed in Subsection 3.3.4. This subsection will also include a brief discussion on the convexity of level sets of positive solutions to the Dirichlet problem (3.48) and (3.49) when the underlying domain Ω is convex.

Finally, we include a brief account for symmetry results to elliptic systems in Subsection 3.3.5.

3.3.1 Symmetry of Semilinear Elliptic Equations in a Ball

In this subsection, we would like to describe various symmetry properties for *positive* solutions of the elliptic equation (3.48) in the case when Ω is a ball of radius R in $\mathbb{R}^n$, under either the homogeneous Dirichlet boundary condition (3.49) or the homogeneous Neumann boundary condition (3.50). Our goal here is to understand what kind of symmetries are being imposed on *all positive* solutions of (3.48) by different boundary conditions (3.49) and (3.50).

Our first result says that the Dirichlet boundary condition (3.49) is "rigid and coercive"—solutions of (3.48) and (3.49) *basically inherit* the symmetries of the domain Ω. (See Subsection 3.3.2 for exceptions and more discussions.)

Theorem 3.13 (see [GNN1]). *Let u be a positive solution of the Dirichlet problem* (3.48) *and* (3.49) *where f is locally Lipschitz continuous. Then u must be radially symmetric; i.e., $u(x) = u(|x|)$, and $u'(r) < 0$ for all $0 < r < R$.*

Observe that there is essentially no condition imposed on f. Therefore, the radial symmetry of solutions to (3.48) and (3.49) seems to result from the symmetry of the domain Ω and the Dirichlet boundary condition. The proof makes use of the well-known "moving plane" method devised by A. D. Alexandroff in 1956.

For simplicity we will sketch the proof of Theorem 3.13 only in the case $0 \le f \in C^1$. The general case can be proved in a similar manner with a little extra work.

Define

$$\Sigma_\lambda = \{x = (x_1, \dots, x_n) \in \Omega \,|\, x_1 > \lambda\},$$

and let T_λ be the hyperplane which is perpendicular to the x_1-axis at $x_1 = \lambda$. Denote the following statement by $(*)_\lambda$:

$$u(x) < u(x^\lambda) \quad \text{for all } x \in \Sigma_\lambda, \quad \text{and } \frac{\partial u}{\partial x_1} < 0 \quad \text{on } \Omega \cap T_\lambda, \qquad (*)_\lambda$$

where x^λ is the reflection of x with respect to T_λ.

We claim that *the set* $\Lambda = \{\lambda \in (0, R) \,|\, (*)_\lambda$ *holds*$\}$ *is nonempty, open, and closed in* $(0, R)$. The fact that $\lambda \in \Lambda$ for all $\lambda < R$ and sufficiently close to R is an easy consequence of the Hopf Boundary Point lemma. The assertion that Λ is both open and closed follows from the following lemma.

Lemma 3.14. *Assume that for some* $\lambda \in (0, R)$ *it holds that* $u(x) \le u(x^\lambda)$ *for all* $x \in \Sigma_\lambda$ *and* $\frac{\partial u}{\partial x_1} \le 0$ *in* Σ_λ. *Then* $(*)_\lambda$ *holds.*

Proof. Let Σ'_λ be the reflection of Σ_λ with respect to T_λ. In Σ'_λ, set $v(x) = u(x^\lambda)$. Then

$$\Delta v + f(v) = 0$$

in Σ'_λ. Considering $w = u - v$ in Σ'_λ, we see that $w \le 0$ in Σ'_λ and $w < 0$ on $\partial \Sigma'_\lambda \backslash T_\lambda$. Thus $w \not\equiv 0$ and

$$\Delta w + c(x) w = 0$$

in Σ'_λ, where

$$c(x) = \begin{cases} \dfrac{f(u(x)) - f(v(x))}{u(x) - v(x)} & \text{if } u(x) \ne v(x), \\ f'(u(x)) & \text{if } u(x) = v(x). \end{cases}$$

Now, $(*)_\lambda$ follows from the Maximum Principle and the Hopf Boundary Point lemma, and our assertion is proved. □

From the assertion we conclude easily that $\Lambda = (0, R)$. Letting $\lambda \to 0, (*)_\lambda$ implies that $u(x) \le u(x^0)$ if $x_1 > 0$. Reversing the x_1-axis, we see that $u(x) = u(x^0)$ if $x_1 > 0$. Since we may replace the x_1-axis by any other direction, Theorem 3.13 is established.

The method above clearly applies to more general domains and equations. In Subsection 3.3.6, we shall include some generalizations of Theorem 3.13.

Naturally, one wonders if Theorem 3.13 would hold for solutions to Neumann boundary value problems (3.48) and (3.50). The results in Section 3.1 clearly show that this is *not* the case. However, solutions to the homogeneous Neumann problem (3.48)–(3.50), although not generally inheriting the full radial symmetry, *do often possess some partial symmetries*. For instance, the "least-energy" solution $u_{\varepsilon,N}$ of (3.1) in Theorem 3.3 was shown to have axial symmetry with the axis of symmetry being the diameter passing through the (single) peak of $u_{\varepsilon,N}$ on $\partial\Omega$. (See [NT2, Section 5].) On the other hand, in [Gr] the single-boundary-peak solution was proved to be unique (in the class of all single-boundary-peak functions) up to rotations. Combining these two results, we see immediately that single-boundary-peak solutions of (3.1) with Ω being a ball and ε small must have axial symmetry. This fact was also established later in [LT] directly. Moreover, it was also shown in [LT] that, for $\varepsilon > 0$ sufficiently small, *all single-peak solutions of* (3.1) *with* Ω *being a ball must either have its peak located at the center of* Ω*, in which case the solution must be radially symmetric, or have its peak located on the boundary, in which case the solution must then be axially symmetric.*

The proof in [LT] is based on a "rotating plane" technique, which is a variant of the "moving plane" method described above.

Pushing the "rotating plane" technique further, Lin, Takagi, and Wei in [LT], [LW] proved, for ε sufficiently small, *all double-peak solutions of* (3.1) *with* Ω *being a ball must be axially symmetric with the axis of symmetry being the diameter passing through the two peaks (and the center of* Ω*).* Moreover, these authors also completely classified, for ε small, all double-peak solutions of (3.1) when $\Omega = B_R(0)$: *For* ε *small, there are only three double-peak solutions of* (3.1) *(up to rotations)—a solution with two boundary peaks situated at* $(-R,0,\dots,0)$ *and* $(R,0,\dots,0)$*, a solution with two interior peaks located at* $(-R/2,0,\dots,0)$ *and* $(R/2,0,\dots,0)$*, and a solution with two mixed peaks at* $(-R,0,\dots,0)$ *and* $(R/3,0,\dots,0)$. In [LW] an attempt was made to classify all *triple-peak* solutions of (3.1) when $\Omega = B_R(0)$. We omit the details here.

Thus, due to the fact that the homogeneous Neumann problems (3.48) and (3.50) tend to allow multiple solutions, we need to "narrow" down specific classes of solutions in studying their symmetry properties.

3.3.2 Symmetry of Semilinear Elliptic Equations in an Annulus

In this subsection, we shall always denote $\Omega = \{x \in \mathbb{R}^n \,|\, 0 < a < |x| < b\}$.

Unlike the case for balls (Theorem 3.13), even solutions to the Dirichlet problem (3.48) and (3.49) do *not* generally have radial symmetry, although annuli also possess radial symmetry. To the best of our knowledge, the first such example was due to Schaeffer [Sc]. Similar ideas were used later by many other authors [BrN], [Co], [DN], [Ly], [NSu]. In some sense a "quantitative" version of such ideas for (3.2) is contained in Theorem 3.4 which, in particular, guarantees that a least-energy solution $u_{\varepsilon,D}$ for (3.2) is *not radially symmetric* in the annulus Ω when ε is small.

However, it is worthwhile to remark that the "moving plane" method does imply that any solution of (3.48) and (3.49) must be decreasing in the radial direction in the "outer" half of Ω, i.e., in $\Omega_o = \{x \in \mathbb{R}^n | \frac{a+b}{2} < |x| < b\}$. Thus, the peak P_ε of a least-energy solution $u_{\varepsilon,D}$ of (3.2) must be in the "inner" half of Ω, and it follows from Theorem 3.4 that P_ε approaches the middle circle $|x| = \frac{a+b}{2}$ as $\varepsilon \to 0$. Finally, we remark that it seems very interesting to compare this result to Theorem 3.11.

3.3.3 Symmetry of Semilinear Elliptic Equations in Entire Space

Systematically exploring symmetry properties of *positive* solutions to semilinear elliptic equations in entire space

$$\begin{cases} \Delta u + f(u) = 0 & \text{in } \mathbb{R}^n, \\ u > 0 & \text{in } \mathbb{R}^n, \\ u \to 0 & \text{at } \infty, \end{cases} \tag{3.51}$$

where $f(0)=0$, also seems to have started in the papers [GNN1, GNN2] around 1980. Much has been done since then. Here, for exposition purposes, we will include only results that are simple but not necessarily most general possible.

It turns out that symmetry results for solutions of (3.51) roughly fall into two categories: $f' \le 0$ near 0, or otherwise. ((3.51) does not have any solution if $f'(0) > 0$.) The case $f' \le 0$ near 0, which includes equations like (3.10), is relatively easier to handle.

Theorem 3.15 (see [LiN]). *Suppose that*

$$f'(s) \le 0 \text{ for sufficiently small } s \ge 0. \tag{3.52}$$

Then all solutions of (3.51) must be radially symmetric about the origin (up to a translation) and $\frac{\partial u}{\partial r} < 0$ for all $r = |x| > 0$.

It was a common belief that the reason for the case $f'(0) < 0$ being easier to handle was that the solution of (3.51) in this case may decay rapidly; in fact, solutions to (3.10) decay exponentially at ∞. However, this turns out to be unnecessary—solutions of (3.51) under hypothesis (3.52) could have very slow decay at ∞. We include such an example here.

Example 3.16 (see [LiN]). *Let m be a positive integer, and let*

$$V_m(x) = \log[\log[\ldots[\log(M+|x|^2)]\ldots]],$$

where V_m is the mth iterated logarithmic function, and M is a positive number such that $V_m(0)=1$. Obviously, V_m is an increasing function of $|x|$ with $\lim_{x\to\infty} V_m(x) = \infty$. Let

$$u_m(x) = \frac{1}{V_m(x)}, \quad x \in \mathbb{R}^2.$$

Then u_m satisfies the following semilinear equation in $\mathbb{R}^2$:

$$\Delta u_m + f_m(u_m) = 0,$$

where

$$\begin{aligned} f_m(s) = &-4s^2 \exp\left(-\sum_{j=1}^{m} \exp^{[j-1]}\left(\frac{1}{s}\right)\right) \\ &\times \left[\left(1 - M\exp\left(-\exp^{[m-1]}\left(\frac{1}{s}\right)\right)\right) \times \left(2s\exp\left(-\sum_{j=1}^{m-1} \exp^{[j-1]}\left(\frac{1}{s}\right)\right)\right.\right. \\ &\left.\left. + \sum_{j=i}^{m-1} \exp\left(-\sum_{j=i}^{m-1} \exp^{[j-1]}\left(\frac{1}{s}\right)\right)\right) - M\exp\left(-\exp^{[m-1]}\left(\frac{1}{s}\right)\right)\right], \end{aligned}$$

and $\exp^{[i]}$ *denotes the ith iterated exponential function, i.e.,*

$$\exp^{[i]}(s) = \exp(\exp(\ldots(\exp(s))\ldots))$$

with $\exp^{[0]} =$ *identity. It can be easily verified that* $f_m(0) = 0$ *and there exists* $r_m > 0$ *such that* $f'_m(s) < 0$ *in* $(0, r_m)$*. Therefore the hypotheses of Theorem* 3.15 *are satisfied. However, as m increases, the decay rate of* u_m *gets slower and slower.*

Proof of Theorem 3.15. In fact, *the novelty of Theorem* 3.15 *is precisely that no decay assumption at* ∞ *is imposed or needed, and the proof is simple.*

As in Subsection 3.3.1, we set $T_\lambda = \{x = (x_1, \ldots, x_n) \in \mathbb{R}^n \,|\, x_1 = \lambda\}$, $\Sigma_\lambda = \{x \in \mathbb{R}^n \,|\, x_1 > \lambda\}$, and we denote the reflection of $x = (x_1, \ldots, x_n)$ with respect to the hyperplane T_λ by $x^\lambda = (2\lambda - x_1, x_2, \ldots, x_n)$. We shall denote the reflection of Σ_λ with respect to T_λ by $\Sigma'_\lambda = \{x \in \mathbb{R}^n \,|\, x_1 < \lambda\}$, as sometimes it is more convenient to deal with Σ'_λ.

Let u be a positive solution of (3.51). First, we define

$$\Lambda = \left\{ \lambda \in \mathbb{R} \,\middle|\, u(x) < u(x^\lambda) \text{ for all } x \in \Sigma_\lambda \text{ and } \frac{\partial u}{\partial x_1} < 0 \text{ on } T_\lambda \right\}.$$

By condition (3.52), there exists $\delta > 0$ such that $f'(s) \leq 0$ for all $0 \leq s \leq \delta$. Since u tends to 0 at ∞, there exist $R_1 > R_0 > \frac{1}{\delta}$ such that

$$\max_{\mathbb{R}^n \setminus B_{R_0}(0)} u < \delta \quad \text{and} \quad \max_{\mathbb{R}^n \setminus B_{R_1}(0)} u < \min_{\overline{B_{R_0}(0)}} u \equiv m_0. \tag{3.53}$$

As the first step in our proof, *we claim that* $[R_1, \infty) \subseteq \Lambda$. To this end, we proceed as follows. For each $\lambda \geq R_1$, let $v(x) = u(x^\lambda)$ and $w(x) = u(x) - v(x)$, for $x \in \Sigma'_\lambda$. Then

$$\Delta w + c(x) w = 0 \text{ in } \Sigma'_\lambda,$$

where

$$c(x) = \begin{cases} \dfrac{f(u(x)) - f(v(x))}{u(x) - v(x)} & \text{if } u(x) \neq v(x), \\ f'(u(x)) & \text{if } u(x) = v(x). \end{cases}$$

Since $\lambda \geq R_1, w > 0$ on $\overline{B_{R_0}(0)}$ by (3.53). On the other hand, from the choice of δ and (3.53) it follows that $c(x) \leq 0$ in $\Sigma'_\lambda \setminus B_{R_0}(0)$. Since $w \geq 0$ on $\partial(\Sigma'_\lambda \setminus \overline{B_{R_0}(0)})$ and $\lim_{x \to \infty} w = 0$ (for $x \in \Sigma'_\lambda$), we conclude from the Maximum Principle and the Hopf Boundary Point lemma that $w > 0$ in $\Sigma'_\lambda \setminus \overline{B_{R_0}(0)}$ and $\frac{\partial w}{\partial x_1} < 0$ on T_λ. Thus $\lambda \in \Lambda$ and our assertion is established.

The rest of the proof proceeds similarly as before and is therefore omitted here. □

Although Theorem 3.15 is already quite general and covers a wide range of equations, the remaining borderline case $f'(0) = 0$ and $f > 0$ in $(0, \delta)$ does include some important examples. For instance, the equation

$$\begin{cases} \Delta u + u^p = 0 & \text{in } \mathbb{R}^n, \\ u > 0 \quad \text{in } \mathbb{R}^n & \text{and} \quad u \to 0 \text{ at } \infty, \end{cases} \tag{3.54}$$

where the exponent $p \geq \frac{n+2}{n-2}, n \geq 3$, has attracted the attention of many mathematicians. All the *radial* solutions of (3.54) have been understood, and they possess remarkable, and perhaps unexpected, properties. (See [Wa1], [Li], [GNW1, GNW2], and [PY].) However, the study of symmetry properties of (3.54) remains a major open problem. Only the critical case of (3.54), where $p = \frac{n+2}{n-2}$, has been resolved.

Theorem 3.17. *All solutions of the problem*

$$\Delta u + u^{\frac{n+2}{n-2}} = 0 \quad \text{in } \mathbb{R}^n \text{ and } u > 0 \text{ in } \mathbb{R}^n \tag{3.55}$$

must take the form

$$u(x) = \left(\frac{\sqrt{n(n-2)\lambda^2}}{\lambda^2 + |x - x_0|^2} \right)^{\frac{n-2}{2}},$$

where $\lambda > 0$ *and* $x_0 \in \mathbb{R}^n$.

Note that *no condition on the asymptotic behavior* of the solution u is imposed in (3.55). We refer readers to [CL] for a brief history and a short, ingenious proof of the remarkable theorem originally due to [CGS].

3.3.4 Related Monotonicity Properties, Level Sets, and More General Domains

The publication of [GNN1] in 1979 has stimulated much research in this direction. In particular, there have been many variants of the "moving plane" method applied to various different domains and/or different types of solutions. (We have encountered one in Subsection 3.3.1 already.) Part of the conclusion resulting from the "moving plane" method is that the solution must be monotone (in addition to being radially symmetric).

In 1991, a useful "sliding" method was devised by Berestycki and Nirenberg [BN]. It was used, for instance, to establish the following result in [BCN1], which deals with more general unbounded domains than just $\mathbb{R}^n$.

Consider the following problem:

$$\begin{cases} \Delta u + f(u) = 0 \quad \text{in } \Omega, \\ u > 0 \quad \text{in } \Omega \quad \text{and} \quad u = 0 \quad \text{on } \partial\Omega, \end{cases} \tag{3.56}$$

where $\Omega = \{x = (x_1, \ldots, x_n) \in \mathbb{R}^n \,|\, x_n > \varphi(x_1, \ldots, x_{n-1})\}$ is an unbounded domain in $\mathbb{R}^n$, $\varphi : \mathbb{R}^{n-1} \to \mathbb{R}$ is a locally Lipschitz continuous function, and f satisfies the following hypothesis:

$$\begin{aligned} &\text{There exist } 0 < s_0 < s_1 < \mu \text{ such that } f(s) \geq \delta_0 s \\ &\text{on } [0, s_0) \text{ for some } \delta_0 > 0, \text{ nonincreasing on } (s_1, \mu), \\ &\text{and } f > 0 \text{ on } (0, \mu), f \leq 0 \text{ on } (\mu, \infty). \end{aligned} \tag{3.57}$$

Theorem 3.18. *Let u be a bounded solution of* (3.56) *with* $M = \sup u < \infty$. *Suppose that* (3.57) *holds. Then u must be monotone in x_n; i.e.,* $\frac{\partial u}{\partial x_n} > 0$ *in* Ω.

In particular, the theorem above applies to domains including half-space. However, in this case, much stronger results for more general $f(u)$ are available. For instance, the following theorem was proved in [BCN1].

Theorem 3.19. *Let u be a bounded solution of*

$$\begin{cases} \Delta u + f(u) = 0 & \text{in } H = \{x = (x_1, \dots, x_n) \in \mathbb{R}^n \,|\, x_n > 0\}, \\ u > 0 \quad \text{in } H \quad \text{and} \quad u = 0 & \text{on } \partial H, \end{cases}$$

where f is locally Lipschitz. If $f(M) \leq 0$, where $M = \sup u$, then u is a function of x_n alone and $\frac{\partial u}{\partial x_n} > 0$ in H.

Incidentally, in [BCN1] it was conjectured that *if there is a solution in Theorem* 3.19, *then necessarily $f(M) = 0$.* This conjecture has been verified only in $n = 2$ by Jang [J] in 2002.

In this connection we ought to mention a well-known conjecture of De Giorgi in 1978.

Conjecture (De Giorgi). *Let u be a solution of*

$$\Delta u + u - u^3 = 0$$

in $\mathbb{R}^n$ with $|u| \leq 1$ and $\frac{\partial u}{\partial x_n} > 0$ in $\mathbb{R}^n$. Then all level sets $[u = \lambda]$ of u are hyperplanes, at least for $n \leq 8$.

This conjecture was proved by Ghoussoub and Gui [GG1] for $n = 2$ in 1998 and by Ambrosio and Cabré [AmC] for $n = 3$ in 2000, and significant progress was made by [GG2] for $n = 4,5$ and the conjecture has been established under an extra condition in [S] for $n \leq 8$ recently. Finally, a counterexample was constructed for $n \geq 9$ in [DKW]. Here we will describe the basic ideas used in [GG1]. In fact, a much more general result was established in [GG1].

Theorem 3.20. *Let $f \in C^1$. Suppose that u is a bounded solution of*

$$\Delta u + f(u) = 0$$

on $\mathbb{R}^2$ with $\frac{\partial u}{\partial x_2} \geq 0$ in $\mathbb{R}^2$. Then u is of the form

$$u(x) = g(ax_1 + bx_2)$$

for some appropriate constants $a, b \in \mathbb{R}$.

This is truly a beautiful theorem. Its proof makes use of the following result of [BCN2].

Proposition 3.21. *Let $L = -\Delta - V$ be a Schrödinger operator on $\mathbb{R}^n$ with the potential V being bounded and continuous. If $Lu = 0$ has a bounded, sign-changing solution, then the first eigenvalue*

$$\lambda_1(V) = \inf\left\{ \frac{\int_{\mathbb{R}^n}(|\nabla\psi|^2 - V\psi^2)}{\int_{\mathbb{R}^n}\psi^2} \,\middle|\, \psi \in C_0^\infty(\mathbb{R}^n) \right\} < 0,$$

provided that $n = 1$ or 2.

Proof of Theorem 3.20. Following [GG1], we now proceed to prove Theorem 3.20. First, we may assume that $\frac{\partial u}{\partial x_2} > 0$ in $\mathbb{R}^2$; otherwise, we will have $\frac{\partial u}{\partial x_2} \equiv 0$ in $\mathbb{R}^2$ by the Maximum Principle and we are done.

Next, observe that $\frac{\partial u}{\partial x_2}$ satisfies the equation

$$\Delta\varphi + V(x)\varphi = 0 \tag{3.58}$$

in $\mathbb{R}^2$, where $V(x) = f'(u(x))$ is bounded and continuous. Since $\frac{\partial u}{\partial x_2} > 0$ in $\mathbb{R}^2$, it follows that $\lambda_1(V) \geq 0$, and (3.58) has no bounded, sign-changing solution (by Proposition 3.21).

On the other hand, given a point $x_0 \in \mathbb{R}^2$, we can choose a direction ν such that $\nu \cdot \nabla u(x_0) = 0$. Since $\partial_\nu u$ also satisfies (3.58), we have $\partial_\nu u \equiv 0$ in $\mathbb{R}^2$, i.e., u is constant along the direction ν and our proof is complete. □

Incidentally, Proposition 3.21 is false for $n \geq 3$. (See [GG1] and [B].)

Concerning properties of the level set of solutions in bounded smooth domains without radial symmetry, some progress has been made as well.

When Ω is convex, it seems natural to ask if the level sets of positive solutions, namely, $\{x \in \Omega \,|\, u(x) \geq \mu\}$, to the Dirichlet problem (3.48) and (3.49) are convex.

Even for the very special case $f(u) = \lambda_1 u$, where λ_1 is the first eigenvalue of $-\Delta$ on Ω under the zero Dirichlet boundary condition, it was a long-standing conjecture that the level sets of the first eigenfunction for a convex domain are convex. This conjecture was proved by Brascamp and Lieb [BL] in 1976 by using the heat equation and log concave functions. Since then, techniques involving Maximum Principles for elliptic equations have been developed by several authors, including Korevaar [Ko], Kennington [Kn], Caffarelli and Friedman [CF], and Korevaar and Lewis [KoL]. The basic idea is to show that, instead of the solution $u, v = g(u)$ is convex for some properly chosen transformation g, which implies that the level sets of v (and therefore those of u) are convex. The transformation g is suitably chosen so that $v = g(u)$ satisfies the equation

$$\Delta v = h(v, \nabla v),$$

where h satisfies

$$h > 0 \quad \text{and} \quad \left(\frac{1}{h}\right)_{vv} \geq 0.$$

Then it was established, in case $n = 2$ by [CF] and $n \geq 3$ by [KoL], that $v = g(u)$ is convex in Ω. The effect of g is to "bend" the graph of u, making it nearly "vertical" near $\partial\Omega$. However, for given f, there seems to be no known general algorithm for finding g.

Recently, there has been some renewed interest in this direction. We refer readers to [MX] for current developments.

Finally, we remark that for Neumann problems, it seems little is known concerning convexity properties of level sets of solutions.

3.3.5 Symmetry of Nonlinear Elliptic Systems

Some of the symmetry results described in previous sections have been generalized to positive solutions of nonlinear elliptic systems. In this subsection, we will mention only two of them: one for balls, the other one for the entire space $\mathbb{R}^n$.

It turns out that Theorem 3.13 can be generalized to *cooperative* elliptic systems in a straightforward manner. (See [Ty].) The elliptic system

$$\begin{cases} \Delta u_i + f_i(u_1,\dots,u_m) = 0 \quad \text{in } \Omega,\ i = 1,\dots,m, \\ u_i > 0 \quad \text{in } \Omega \quad \text{and} \quad u_i = 0 \quad \text{on } \partial\Omega, \end{cases} \tag{3.59}$$

is said to be *cooperative* if f_i is C^1 and

$$\frac{\partial f_i}{\partial u_j} \geq 0 \quad \text{for all } i \neq j \text{ and } 1 \leq i, j \leq m. \tag{3.60}$$

Theorem 3.22. *Let Ω be a ball of radius R in $\mathbb{R}^n$, let f satisfy* (3.60), *and let $(u_1,\dots,u_m)$ be a solution of* (3.59). *Then, for each i, u_i is radially symmetric and $u_i'(r) < 0$ for $0 < r = |x| < R$.*

Since the usual Maximum Principle for single elliptic equations generalizes to *cooperative elliptic systems* [PrW], the proof of Theorem 3.13 also generalizes naturally to establish Theorem 3.22.

The second result here generalizes Theorem 3.15 for the entire space. This is more involved. Here we are dealing with solutions of the following problems:

$$\begin{cases} \Delta u_i + f_i(u_i,\dots,u_m) = 0 \quad \text{in } \mathbb{R}^n,\ i = 1,\dots,m, \\ u_i > 0 \quad \text{in } \mathbb{R}^n \quad \text{and} \quad u_i(x) \to 0 \quad \text{as } x \to \infty. \end{cases} \tag{3.61}$$

In addition to (3.60), we will also assume that $f_i, i = 1,\dots,m$, satisfy the following hypotheses:

$$\begin{aligned} &\text{There exists } \varepsilon > 0 \text{ such that the system (3.61) is \textit{fully coupled}} \\ &\text{in } 0 < u < \varepsilon; \text{ more precisely, for any } I, J \subseteq \{1,\dots,m\} \text{ with} \\ &I \cap J = \phi \text{ and } I \cup J = \{1,\dots,m\}, \text{ there exist } i_0 \in I \text{ and } j_0 \in J \\ &\text{such that } \frac{\partial f_{i_0}}{\partial u_{j_0}} > 0 \text{ in } 0 < u < \varepsilon. \end{aligned} \tag{3.62}$$

$$\begin{aligned} &\text{All \textit{principal minors} of } -A(u_1,\dots,U_m) \text{ have nonnegative} \\ &\text{determinants for } 0 < u < \varepsilon, \text{ where} \\ &A(u_1,\dots,u_m) = \left(\frac{\partial f_i}{\partial u_j}\right)_{1 \leq i,j \leq m}. \end{aligned} \tag{3.63}$$

Recall that the principal minors of a matrix $(m_{ij})_{1 \leq i,j \leq m}$ are the submatrices

$$(m_{ij})_{1 \leq i,j \leq k}, \quad 1 \leq k \leq m.$$

Observe that (3.62) is to guarantee that all $u_i, i = 1,\dots,m$, are radially symmetric with respect to the same point, while (3.63) reduces to (3.52) in Theorem 3.15 in the single equation case.

In [BS] the following result is proved.

Theorem 3.23. *Suppose that* (3.60), (3.62), *and* (3.63) *hold and* u *is a solution of* (3.61). *Then* u *must be radially symmetric (up to a translation), and* $u'(r) < 0$ *for* $r = |x| > 0$.

The proof, still using the "moving plane" method, is more involved. We refer interested readers to [BS].

3.3.6 Generalizations, Miscellaneous Results, and Concluding Remarks

So far in this section we have focused our attention on "autonomous" equations or systems with pure diffusion (i.e., Laplacians); here in this subsection we shall briefly discuss some generalizations and related results.

Generally speaking, if we replace the term $f(u)$ in (3.48) by $f(r,u) = f(|x|,u)$, then symmetry results Theorems 3.13 and 3.15 still hold, provided that $f(r,u)$ is nonincreasing in $r > 0$. On the other hand, if $f(r,u)$ is increasing in r, one cannot expect solutions to be radially symmetric anymore. For instance, the Dirichlet problem

$$\begin{cases} \Delta u - u + V(|x|)u^P = 0 & \text{in } B_R \subseteq \mathbb{R}^n, \\ u > 0 \quad \text{in } B_R \quad \text{and} \quad u = 0 & \text{on } \partial B_R \end{cases}$$

has nonradially symmetric solutions for R large, where $p > 1, V(|x|) = 1 + |x|^\ell$, and $0 < \ell < \frac{(n-1)(p-1)}{2}$. (See, e.g., [DN, Proposition 5.10.]) Similarly, so does its counterpart for entire space.

In [TI], the following problem was considered:

$$\begin{cases} \Delta u + f(|x|,u) = 0 & \text{in } \mathbb{R}^n,\ n \geq 3, \\ u \to \infty & \text{at } \infty. \end{cases} \tag{3.64}$$

Under the assumption that f, roughly speaking, is monotone in u with superlinear growth in u when $|x|$ and u both are large and positive, and $r^{2n-2}f(r,u)$ is "asymptotically" monotone in r for u large, it is established in [TI] that *all solutions of* (3.64) *are radially symmetric*. The proof consists of two parts: First, prove that the difference of any two solutions of (3.64) must tend to 0 as $|x| \to \infty$; then apply the arguments of [LiN] described in Subsection 3.3.3.

In case $n = 2$, the method in [TI] yields a similar result with the "boundary value $u \to \infty$ as $|x| \to \infty$" in (3.64) replaced by

$$\frac{u(x)}{\log|x|} \to \infty \quad \text{as } |x| \to \infty \tag{3.65}$$

and with another technical condition imposed on the monotonicity of f with respect to r. It is curious to note that there is an earlier result, due to [CN], asserting that *all solutions of* (3.64) *in* $\mathbb{R}^2$ *with*

$$f(r,u) = K(r)e^u, \tag{3.66}$$

where $K \geq 0$ *and* $K \sim |x|^{-\ell}$ *at* ∞, *for some* $\ell > 2$, *are radially symmetric. In fact, in this case all solutions are completely understood and classified; in particular, there is no solution having the asymptotic behavior* (3.65). Incidentally, the equation in (3.64) with the nonlinearity (3.66) is known as the *conformal Gaussian curvature equation* with K as the prescribed Gaussian curvature in $\mathbb{R}^2$.

One can also replace the Laplace operator in (3.48) by other operators, e.g., by fully nonlinear operators $F(x,u(x),Du(x),D^2u(x))=0$, where F satisfies the following:

(F1) $F(x,s,p_i,p_{ij}), 1\le i,j\le n$, is continuous in all of its variables, C^1 in p_{ij}, and Lipschitz in s and p_i, where p_{ij}'s are position variables for $\frac{\partial^2 u}{\partial x_i \partial x_j}$, p_i for $\frac{\partial u}{\partial x_i}$ and s for u;

(F2) $F_{p_{ij}}(x,s,p_i,p_{ij})\xi_i\xi_j \ge \overline{\lambda}(s,x,p_i,p_{ij})\|\xi\|^2$ for all $\xi\in\mathbb{R}^n$, where $\overline{\lambda}>0$ in $\mathbb{R}^n\times\mathbb{R}\times\mathbb{R}^n\times\mathbb{R}^{n^2}$;

(F3) $F(x,s,p_i,p_{ij}) = F(|x|,s,p_i,p_{ij})$ and F is nonincreasing in $|x|$;

(F4) $F(x,s,p_1,\ldots,p_{i_0-1},-p_{i_0},p_{i_0+1},\ldots,p_n,p_{11},\ldots,\quad -p_{i_0j_0},\ldots,-p_{j_0i_0},\quad \ldots,p_{nn}) = F(x,s,p_i,p_{ij})$ for $1\le i_0,j_0\le n$ and $i_0\ne j_0$.

Theorem 3.24 (see [LiN]). *Suppose that F satisfies* (F1)–(F4) *and that $F_s\le 0$ for $|x|$ large and for s small and positive. Let u be a positive C^2 solution of*

$$\begin{cases} F\left(x,u(x),Du(x),D^2u(x)\right)=0 & \text{in } \mathbb{R}^n, \quad n\ge 2,\\ u(x)\to 0 & \text{at } \infty. \end{cases}$$

Then u must be radially symmetric (up to a translation) and $u_r<0$ for $r=|x|>0$.

The proof uses essentially the same arguments as in that of Theorem 3.15 and thus is omitted here. However, we wish to remark here that the elliptic operator F in Theorem 3.24 is *not* required to be *uniformly elliptic* and therefore is quite general. For instance, it includes the minimal surface operator, or equations of mean-curvature type,

$$\begin{cases} \operatorname{div}\left(\frac{Du}{\sqrt{1+|Du|^2}}\right)+f(u)=0 \quad \text{in } \mathbb{R}^n,\\ u>0 \quad \text{in } \mathbb{R}^n \quad \text{and} \quad u\to 0 \quad \text{at } \infty. \end{cases} \tag{3.67}$$

Consequently, Theorem 3.24 also contains previous work [FL] on (3.67).

In this direction we ought to discuss the p-Laplacian

$$\Delta_p u = \operatorname{div}(|Du|^{p-2}Du),$$

where $p>1$, which exhibits certain degeneracy or singularity depending on $p>2$ or $p<2$. (Note that the case $p=2$ gives rise to the usual Laplace operator.) The case $1<p<2$ is studied in [DPR]. It is proved there that *essentially under the same hypothesis* (3.52) *a solution u of the problem*

$$\begin{cases} \Delta_p u + f(u)=0 \quad \text{in } \mathbb{R}^n,\\ u>0 \quad \text{in } \mathbb{R}^n, \quad u\to 0 \quad \text{at } \infty, \end{cases} \tag{3.68}$$

where $1<p<2$, must be radially symmetric (up to a translation) and $u_r<0$ in $r=|x|>0$.

The method of proof in [DPR] still uses the "moving plane" technique, but with a weak comparison principle instead of the usual Maximum Principle.

The symmetry of solutions to (3.68) for the degenerate case $p > 2$ does not hold in general, however. See [SZ, Section 6] for a counterexample.

Chapter 4

Diffusion in Heterogeneous Environments: 2 x 2 Lotka–Volterra Competition Systems

In Chapter 2 we touched upon the interaction between diffusion and spatial heterogeneity in general 2×2 reaction-diffusion systems and their shadow systems. Here in this chapter we shall explore further in this direction with some depth using the classical Lotka–Volterra competition-diffusion systems as our main example in illustrating various interesting aspects of this interaction.

Suppose that we have two competing species, U and V, with different (random) dispersal rates but that are otherwise identical. If the habitat Ω is spatially *inhomogeneous*, then *the slower diffuser always prevails*! That is, the slower diffuser always wipes out its faster competitor, regardless of the initial population densities (as long as both are nonzero).

This interesting result was proved in 1998 by [DHMP]. Mathematically, it may be described as follows. Suppose that the species U and V satisfy the classical Lotka–Volterra competition-diffusion system

$$\begin{cases} U_t = d_1 \Delta U + U(m(x) - U - V) & \text{in } \Omega \times (0,\infty), \\ V_t = d_2 \Delta V + V(m(x) - U - V) & \text{in } \Omega \times (0,\infty), \\ \partial_\nu U = 0 = \partial_\nu V & \text{on } \partial\Omega \times (0,\infty), \end{cases} \tag{4.1}$$

where $0 \le m(x) \not\equiv$ Constant reflects the carrying capacity or the resources at location x in Ω. (As U, V represent the population densities of two competing species, we will always assume that they are nonnegative.) If $d_1 < d_2$, then, with any initial values $U(x,0) \ge 0, \not\equiv 0$ and $V(x,0) \ge 0, \not\equiv 0$, it always holds that the solution of (4.1)

$$(U(x,t),\ V(x,t)) \to (\theta_{d_1}(x),\ 0)$$

uniformly as $t \to \infty$, where θ_{d_1} is the *unique positive steady state* of the problem

$$\begin{cases} u_t = d_1 \Delta u + u(m(x) - u) & \text{in } \Omega \times (0,\infty), \\ \partial_\nu u = 0 & \text{on } \partial\Omega \times (0,\infty); \end{cases} \tag{4.2}$$

in other words, U always wipes out V! (See Section 4.2.) This innocent-looking single equation (4.2) turns out to be important and is interesting in its own right. It is known that for every $d_1 > 0$, (4.2) has a *unique positive steady state* θ_{d_1} *which is globally asymptotically stable.* (See Theorem 4.1.) It seems interesting to observe that if we divide the equation for θ_{d_1} by θ_{d_1}, and then integrate, we have

$$d_1 \int_\Omega \frac{|\nabla \theta_{d_1}|^2}{\theta_{d_1}^2} + \int_\Omega (m(x) - \theta_{d_1}) = 0.$$

Since $m \not\equiv$ Constant, which implies that $\theta_{d_1} \not\equiv$ Constant, we have, *for every* $d_1 > 0$,

$$\int_\Omega m(x) < \int_\Omega \theta_{d_1}. \tag{4.3}$$

In other words, *with diffusion, the total population is always greater than the total carrying capacity!* Moreover, it is well known that

$$\theta_{d_1}(x) \to \begin{cases} m(x) & \text{as } d_1 \to 0, \\ \overline{m} \equiv \dfrac{1}{|\Omega|} \displaystyle\int_\Omega m & \text{as } d_1 \to \infty. \end{cases}$$

Thus,

$$\int_\Omega \theta_{d_1} \to \int_\Omega m$$

as $d_1 \to 0$ or ∞.

It seems surprising that we actually know very little about θ_{d_1}, or $\int_\Omega \theta_{d_1}$. For instance, *what is the maximum of* $\int_\Omega \theta_{d_1}$ *for* $d_1 \in (0, \infty)$? *Where is the maximum of* $\int_\Omega \theta_{d_1}$ *assumed?* Moreover, it turns out that the inequality (4.3) plays a key role in the understanding of the phenomenon "slower diffuser always prevails!"

One of the purposes of this chapter is to develop/describe mathematical theories to understand these interesting and curious results. We will follow the approach initiated by Lou in [L1]. After treating a slightly more general case than (4.1), it seems reasonable to question whether slower diffuser would fare better even just in a slightly broader context. (See Section 4.4.) We should mention that the monograph [CC] provides an excellent overview of the background materials for interested readers.

4.1 The Logistic Equation in Heterogeneous Environment

In this section, we include necessary background materials for the (single) logistic equation (4.2), which plays an important role in the rest of this chapter.

It turns out that we need to consider a slightly more general equation than (4.2); namely,

$$\begin{cases} u_t = d\Delta u + u(h(x) - u) & \text{in } \Omega \times (0, \infty), \\ \partial_\nu u = 0 & \text{on } \partial\Omega \times (0, \infty), \end{cases} \tag{4.4}$$

where $h \not\equiv$ Constant, could change sign.

Theorem 4.1. *Suppose that $h \not\equiv$ Constant and h is positive somewhere in Ω.*

(i) *If $\int_\Omega h \geq 0$, then for each $d > 0$ (4.4) has a unique positive steady state $\theta_{d,h}$ which is globally asymptotically stable. Moreover,*

$$\theta_{d,h}(x) \to \begin{cases} h_+(x) & \text{as } d \to 0, \\ \bar{h} \equiv \frac{1}{|\Omega|}\int_\Omega h & \text{as } d \to \infty, \end{cases}$$

where $h_+(x) = \max\{h(x), 0\}$.

(ii) *If $\int_\Omega h < 0$, then (4.4) has a unique positive steady state $\theta_{d,h}$ if and only if $0 < d < 1/\lambda_1(h)$, where $\lambda_1(h)$ is given in* Theorem 4.2. *Moreover, $\theta_{d,h}$ is globally asymptotically stable and $\theta_{d,h} \to h_+$ as $d \to 0$.*

(iii) *If $\int_\Omega h < 0$ and $d \geq 1/\lambda_1(h)$, then all solutions of (4.4) (with initial value $u(x,0) \geq 0$ in Ω) converge to 0 as $t \to \infty$.*

Theorem 4.1 is well known. However, the analysis involved in its proof will be needed in our considerations of the competition system later. Thus, we are including an outline of the proof.

It will be useful to consider that the eigenvalue problem with indefinite weight

$$\begin{cases} \Delta\varphi + \lambda h(x)\varphi = 0 & \text{in } \Omega, \\ \partial_\nu \varphi = 0 & \text{on } \partial\Omega, \end{cases} \tag{4.5}$$

where $h \not\equiv$ Constant, could change sign in Ω. We say that λ *is a principal eigenvalue if* (4.5) *has a positive solution.* (Notice that 0 is always a principal eigenvalue.) The following result is standard.

Theorem 4.2. *The problem* (4.5) *has a nonzero principal eigenvalue $\lambda_1 = \lambda_1(h)$ if and only if h changes sign and $\int_\Omega h \neq 0$. More precisely,*

(i) $\int_\Omega h < 0 \Rightarrow \lambda_1(h) > 0$;

(ii) $\int_\Omega h > 0 \Rightarrow \lambda_1(h) < 0$;

(iii) $\int_\Omega h = 0 \Rightarrow$ *0 is the only principal eigenvalue.*

Moreover, for $\int_\Omega h < 0$, $\lambda_1(h)$ is given by the following variational characterization:

$$\lambda_1(h) = \inf\left\{ \frac{\int_\Omega |\nabla\varphi|^2}{\int_\Omega h\varphi^2} \,\middle|\, \varphi \in H_1(\Omega) \text{ and } \int_\Omega h\varphi^2 > 0 \right\},$$

and

(iv) *$\lambda_1(h) > \lambda_1(k)$ if $h \leq k$ and $h \not\equiv k$;*

(v) *$\lambda_1(h)$ is continuous in h; more precisely, $\lambda_1(h_l) \to \lambda_1(h)$ if $h_l \to h$ in $L^\infty(\Omega)$.*

It turns out that (4.5) is very much related to the following eigenvalue problem:

$$\begin{cases} \Delta\psi + \lambda h(x)\psi + \widetilde{\mu}\psi = 0 & \text{in } \Omega, \\ \partial_\nu \psi = 0 & \text{on } \partial\Omega. \end{cases} \tag{4.6}$$

It is easy to see that *the first eigenvalue of* (4.6), $\widetilde{\mu}_1 = \widetilde{\mu}_1(\lambda, h)$, *is* 0 *if and only* λ *is a principal eigenvalue of* (4.5). Furthermore, it is straightforward to verify that $\widetilde{\mu}_1(\lambda, h)$ *is concave in* λ; i.e.,

$$\begin{aligned} &\widetilde{\mu}_1(\alpha\lambda + (1-\alpha)\bar{\lambda}) \\ &= \inf\left\{ \int_\Omega \left[|\nabla\psi|^2 - (\alpha\lambda + (1-\alpha)\bar{\lambda})h\psi^2 \right] \middle| \int_\Omega \psi^2 = 1, \psi \in H_1(\Omega) \right\} \\ &= \inf\left\{ \alpha \int_\Omega \left[|\nabla\psi|^2 - \lambda h\psi^2 \right] + (1-\alpha)\int_\Omega \left[|\nabla\psi|^2 - \bar{\lambda}h\psi^2 \right] \middle| \int_\Omega \psi^2 = 1, \psi \in H_1(\Omega) \right\} \\ &> \alpha\widetilde{\mu}_1(\lambda) + (1-\alpha)\widetilde{\mu}_1(\bar{\lambda}), \end{aligned}$$

where the last inequality is *strict*, as the corresponding eigenfunctions are different. Thus the following corollary is immediate.

Corollary 4.3. *The first eigenvalue* $\widetilde{\mu}_1(\lambda)$ *of* (4.6) *has the following properties (see Figure* 4.1*):*

(i) $\int_\Omega h \geq 0 \quad \Rightarrow \quad \widetilde{\mu}_1(\lambda) < 0$ *for all* $\lambda > 0$;

(ii)

$$\int_\Omega h < 0 \quad \Rightarrow \quad \begin{cases} \widetilde{\mu}_1(\lambda) < 0 & \text{if } \lambda > \lambda_1(h), \\ \widetilde{\mu}_1(\lambda) = 0 & \text{if } \lambda = \lambda_1(h), \\ \widetilde{\mu}_1(\lambda) > 0 & \text{if } 0 < \lambda < \lambda_1(h). \end{cases}$$

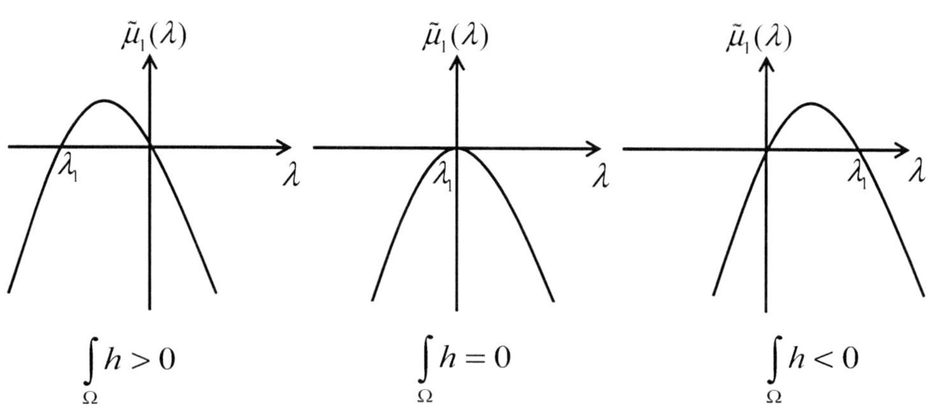

Figure 4.1.

For our purposes, it is more convenient to consider the following eigenvalue problem instead:

$$\begin{cases} d\Delta\psi + h(x)\psi + \mu\psi = 0 & \text{in } \Omega, \\ \partial_\nu \psi = 0 & \text{on } \partial\Omega. \end{cases} \tag{4.7}$$

Thus, the first eigenvalue of (4.7) is given by

$$\mu_1(d,h) = d\widetilde{\mu}_1\left(\frac{1}{d}, h\right). \tag{4.8}$$

From Corollary 4.3 and the variational characterization of $\mu_1(d,h)$, namely,

$$\mu_1(d,h) = \inf\left\{\int_\Omega \left(d|\nabla\psi|^2 - h(x)\psi^2\right)dx \,\middle|\, \int_\Omega \psi^2 = 1,\ \psi \in H_1(\Omega)\right\}, \tag{4.9}$$

we have the following result.

Proposition 4.4. *The first eigenvalue $\mu_1(d,h)$ of (4.7) has the following properties:*

(i) $\int_\Omega h \geq 0 \quad \Rightarrow \quad \mu_1(d,h) < 0 \quad$ *for all* $d > 0$.

(ii)

$$\int_\Omega h < 0 \quad \Rightarrow \quad \begin{cases} \mu_1(d,h) < 0 & \text{for all } d < 1/\lambda_1(h), \\ \mu_1(d,h) = 0 & \text{for all } d = 1/\lambda_1(h), \\ \mu_1(d,h) > 0 & \text{for all } d > 1/\lambda_1(h). \end{cases}$$

(iii) *$\mu_1(d,h)$ is strictly increasing and concave in $d > 0$. Moreover,*

$$\lim_{d\to 0}\mu_1(d,h) = \min_{\overline{\Omega}}(-h) \quad \text{and} \quad \lim_{d\to\infty}\mu_1(d,h) = -\overline{h},$$

where $\overline{h}$ is the average of h.

(iv) *$\mu_1(d,h) < \mu_1(d,k)$ if $h \geq k$ and $h \not\equiv k$.*

The concavity of $\mu_1(d,h)$ follows from the formula (with h fixed)

$$\mu_1''(d) = \frac{1}{d^3}\widetilde{\mu}_1''\left(\frac{1}{d}\right),$$

and the limiting behavior of $\mu_1(d,h)$ can be established by standard arguments.

From Proposition 4.4 the existence and uniqueness of $\theta_{d,h}$ in Theorem 4.1 follow readily by the method of upper and lower solutions [Sa], for any large constant M serves as an upper solution for the steady state equation of (4.4). For a lower solution, we set $\underline{u} = \varepsilon\psi_1$, where ψ_1 is the first eigenfunction of (4.7), and $\varepsilon > 0$ is a small constant. Then

$$d\Delta\underline{u} + \underline{u}(h(x) - \underline{u}) = \varepsilon\psi_1[-\mu_1(d) - \varepsilon\psi_1] > 0$$

as long as $\mu_1 < 0$. This guarantees the existence of $\theta_{d,h}$. The uniqueness of $\theta_{d,h}$ also follows in a standard fashion. Suppose that (4.4) has two positive solutions u_1 and u_2. If $u_1 \le u_2$ in Ω, we have

$$\begin{aligned} 0 &= \int_\Omega [d\Delta u_1 + u_1(h-u_1)]\, u_2 \\ &= \int_\Omega u_1 [d\Delta u_2 + u_2(h-u_1)] \\ &= \int_\Omega u_1 u_2 [-(h-u_2)+(h-u_1)] \\ &= \int_\Omega u_1 u_2 (u_2 - u_1) > 0 \end{aligned}$$

since $u_1 \not\equiv u_2$. Now, if neither $u_1 \le u_2$ nor $u_2 \le u_1$ in Ω, then setting

$$\widetilde{u}(x) = \max\{u_1(x), u_2(x)\},$$

we see that $\widetilde{u}$ is a lower solution. Since any large constant M is always an upper solution, we have a solution $\widetilde{u} \le u_3 \le M$. Thus we have a third solution $u_3 \ge u_1$ and $u_3 \not\equiv u_1$, a contradiction, and uniqueness is proved.

From the uniqueness and the construction of upper and lower solutions, the global asymptotic stability of $\theta_{d,h}$ now follows.

In the rest of this book, when there is no confusion, sometimes we will suppress the indices h, and/or d, in $\theta_{d,h}$ and $\mu_1(d,h)$.

4.2 Slower Diffuser versus Fast Diffuser

We now return to the interesting phenomenon "Slower diffuser always prevails!" described at the beginning of this chapter, where we assumed two competing species U and V, with different dispersal rates but otherwise being identical, satisfy the classical Lotka–Volterra competition-diffusion system (4.1)

$$\begin{cases} U_t = d_1 \Delta U + U(m(x) - U - V) & \text{in } \Omega \times (0,\infty), \\ V_t = d_2 \Delta V + V(m(x) - U - V) & \text{in } \Omega \times (0,\infty), \\ \partial_\nu U = \partial_\nu V = 0 & \text{on } \partial\Omega \times (0,\infty), \end{cases} \tag{4.1}$$

where $m(x) \not\equiv$ Constant and $d_1 < d_2$. Then, regardless of the initial value $U(x,0) \not\equiv 0, V(x,0) \not\equiv 0$, U always wipes out V. Mathematically, we have the following.

Theorem 4.5 (see [DHMP]). *If $d_1 < d_2$ in (4.1), then the steady state $(\theta_{d_1}, 0)$ is globally asymptotically stable, as long as $U(x,0) \ge 0, \not\equiv 0$ and $V(x,0) \ge 0, \not\equiv 0$ in Ω.*

(In this section, we will write θ_{d_1} and θ_{d_2}, instead of $\theta_{d_1,m}$ and $\theta_{d_2,m}$.)

The proof consists of three steps:

Step 1. *$(\theta_{d_1}, 0)$ is locally asymptotically stable and $(0, \theta_{d_2})$ is unstable.*

Step 2. (4.1) *does not have any coexistence steady state* $(\widetilde{U},\widetilde{V})$*; i.e.,* $\widetilde{U}>0$ *and* $\widetilde{V}>0$.

Step 3. *Conclude from the theory of monotone flows that* $(\theta_{d_1},0)$ *is globally asymptotically stable.*

To prove the stability of $(\theta_{d_1},0)$, we linearize the corresponding elliptic system of (4.1) at $(\theta_{d_1},0)$:

$$\begin{cases} d_1\Delta\Psi_1+(m(x)-2\theta_{d_1})\Psi_1-\theta_{d_1}\Psi_2+\lambda\Psi_1=0 & \text{in } \Omega,\\ d_2\Delta\Psi_2+(m(x)-\theta_{d_1})\Psi_2+\lambda\Psi_2=0 & \text{in } \Omega,\\ \partial_\nu\Psi_1=\partial_\nu\Psi_2=0 & \text{on } \partial\Omega. \end{cases} \qquad (L_1)$$

First, observe that all eigenvalues of (L_1) are real since they are also eigenvalues of the second equation in (L_1). Next, note that $\mu_1(d_1,m(x)-\theta_{d_1})=0$. From Proposition 4.4(iii), we have

$$\mu_1(d_2,m(x)-\theta_{d_1})>\mu_1(d_1,m(x)-\theta_{d_1})=0$$

since $d_1<d_2$. Thus, if $\Psi_2\not\equiv 0$, then the corresponding eigenvalue λ of (L_1) must satisfy

$$\lambda\geq\mu_1(d_2,m(x)-\theta_{d_1})>0.$$

If $\Psi_2\equiv 0$, then the corresponding eigenvalue λ is an eigenvalue of the first equation, which reduces to

$$\begin{cases} d_1\Delta\Psi_1+(m(x)-2\theta_{d_1})\Psi_1+\lambda\Psi_1=0 & \text{in } \Omega,\\ \partial_\nu\Psi_1=0 & \text{on } \partial\Omega. \end{cases}$$

Again, from Proposition 4.4(iv), it follows that

$$\lambda\geq\mu_1(d_1,m(x)-2\theta_{d_1})>\mu_1(d_1,m(x)-\theta_{d_1})=0.$$

Therefore all eigenvalues of (L_1) are positive, and $(\theta_{d_1},0)$ is locally asymptotically stable.

To see that $(0,\theta_{d_2})$ is unstable, we also linearize the corresponding elliptic system of (4.1) at $(0,\theta_{d_2})$:

$$\begin{cases} d_1\Delta\Psi_1+(m(x)-\theta_{d_2})\Psi_1+\lambda\Psi_1=0 & \text{in } \Omega,\\ d_2\Delta\Psi_2+(m(x)-2\theta_{d_2})\Psi_2-\theta_{d_2}\Psi_1+\lambda\Psi_2=0 & \text{in } \Omega,\\ \partial_\nu\Psi_1=\partial_\nu\Psi_2=0 & \text{on } \partial\Omega. \end{cases} \qquad (L_2)$$

Similarly, all eigenvalues of (L_2) are real. Setting

$$\lambda=\mu_1(d_1,\, m(x)-\theta_{d_2}),$$

we see that $\lambda<0$ by Proposition 4.4(iii) since $\mu_1(d_2,m(x)-\theta_{d_2})=0$. Now, we let Ψ_1 be the first eigenfunction corresponding to $\mu_1(d_1,m(x)-\theta_{d_2})$, and we proceed to solve Ψ_2 from the second equation in (L_2). But this follows from the fact that the operator

$$d_2\Delta+(m(x)-2\theta_{d_2})+\lambda$$

is invertible since $\lambda < 0$ and

$$\mu_1(d_2, m(x) - 2\theta_{d_2}) > \mu_1(d_2, m(x) - \theta_{d_2}) = 0.$$

Thus (L_2) has a negative eigenvalue, and $(0, \theta_{d_2})$ is unstable. This completes the proof of Step 1.

Step 2, namely, no coexistence steady state, follows easily as follows. Suppose $(\widetilde{U}, \widetilde{V})$ is a coexistence steady state of (4.1), where $\widetilde{U} > 0$ and $\widetilde{V} > 0$. Then, if $\widetilde{U} + \widetilde{V} \not\equiv m(x)$, we have

$$\mu_1(d_1, m(x) - \widetilde{U} - \widetilde{V}) = 0 = \mu_1(d_2, m(x) - \widetilde{U} - \widetilde{V}),$$

a contradiction. If $\widetilde{U} + \widetilde{V} \equiv m(x)$, then we have

$$d_1 \Delta \widetilde{U} \equiv 0 \equiv d_2 \Delta \widetilde{V},$$

which implies that $\widetilde{U} \equiv$ Constant and $\widetilde{V} \equiv$ Constant since $\partial_\nu \widetilde{U} = 0 = \partial_\nu \widetilde{V}$ on $\partial\Omega$. Consequently, $m(x) \equiv \widetilde{U} + \widetilde{V} \equiv$ Constant, again a contradiction.

Now, the theory of monotone flow [H], [He] guarantees that there is a connecting orbit from $(0, \theta_{d_2})$ to $(\theta_{d_1}, 0)$; in particular, $(\theta_{d_1}, 0)$ is globally asymptotically stable, and the proof of Theorem 4.5 is complete. □

It is interesting to note that the proofs of Step 1 and 2 readily generalize to $N \times N$ systems, $N \geq 3$. However, to apply the theory of monotone flow in Step 3, requires that $N = 2$. Thus, when we have N competing species $U_1, \ldots, U_N$, $N \geq 3$, with different dispersal rates $d_1 < d_2 < \cdots < d_N$ but being identical otherwise, *it is not known if the slowest diffuser U_1 would always prevail!*

In the rest of this chapter, following the approach initiated by Lou [L1], we will try to understand this interesting phenomenon by approximating the system (4.1) using Lotka–Volterra weak competition systems. Our main results in this direction are Theorems 4.9 and 4.10 in Section 4.4. The arguments in Section 4.4 seem to indicate that the phenomenon "slower diffuser always prevails!" described above is highly special—in fact, slower diffuser does not always prevail even in a slightly more general context. (See Section 4.4.)

4.3 Lotka–Volterra Competition-Diffusion System in Homogeneous Environment

To provide the background we begin with a brief summary of the Lotka–Volterra competition-diffusion systems with *constant* coefficients (i.e., homogeneous environments)

$$\begin{cases} U_t = d_1 \Delta U + U(a_1 - b_1 U - c_1 V) & \text{in } \Omega \times (0, \infty), \\ V_t = d_2 \Delta V + V(a_2 - b_2 U - c_2 V) & \text{in } \Omega \times (0, \infty), \\ \partial_\nu U = 0 = \partial_\nu V & \text{on } \partial\Omega \times (0, \infty), \end{cases} \tag{4.10}$$

where U, V again represent the population densities of two competing species; the constants a_1, a_2 reflect the carrying capacities of the habitat Ω, b_1, c_2 are the intraspecific competition coefficients, and b_2, c_1 are the interspecific competition coefficients, and they are all positive. We will consider only nonnegative solutions, and it is convenient to denote

$$A = \frac{a_1}{a_2}, \quad B = \frac{b_1}{b_2}, \quad \text{and} \quad C = \frac{c_1}{c_2}.$$

For (4.10) the dynamics of the *weak competition case* $B > A > C$ is particularly simple and is completely understood: In this case (4.10) *has a coexistence (i.e., positive) steady state*

$$(U_*, V_*) = \left(\frac{a_1c_2 - a_2c_1}{b_1c_2 - b_2c_1}, \frac{b_1a_2 - b_2a_1}{b_1c_2 - b_2c_1} \right) \tag{4.11}$$

which is globally asymptotically stable; i.e., regardless of the initial values $U(x,0)$, $V(x,0)$ (as long as $U(x,0) \not\equiv 0$, and $V(x,0) \not\equiv 0$), the solution of (4.10) always converges to (U_*, V_*) as $t \to \infty$ for any diffusion rates d_1, d_2.

Proposition 4.6. *For $B > A > C$, the coexistence constant steady state (U_*, V_*) given by* (4.11) *is globally asymptotically stable.*

Proof. For $U(x,0) \geq 0$, $\not\equiv 0$ and $V(x,0) \geq 0$, $\not\equiv 0$, it follows immediately from the Maximum Principle that $U(x,t) > 0$ on $\overline{\Omega}$ and $v(x,t) > 0$ on $\overline{\Omega}$ for all $t > 0$.

Setting

$$E[U,V](t) = \int_\Omega \left(b_2 \left[U - U_* - U_* \log \frac{U}{U_*} \right] + c_1 \left[V - V_* - \log \frac{V}{V_*} \right] \right) dx$$

we have

$$\begin{aligned}
&\frac{d}{dt} E[U,V](t) \\
&= - \int_\Omega \left[b_1b_2\widehat{U}^2 + 2b_2c_1\widehat{U}\widehat{V} + c_1c_2\widehat{V}^2 \right] - \int_\Omega \frac{d_1b_2U_*}{U^2} |\nabla \widehat{U}|^2 - \int_\Omega \frac{d_2c_1V_*}{V^2} |\nabla \widehat{V}|^2 \\
&\leq - \frac{b_1b_2c_1c_2 - b_2^2c_1^2}{2b_1b_2} \int_\Omega \widehat{V}^2 - \frac{b_1b_2c_1c_2 - b_2^2c_1^2}{2c_1c_2} \int_\Omega \widehat{U}^2 \\
&\quad - \frac{d_1b_2U_*}{M^2} \int_\Omega |\nabla \widehat{U}|^2 - \frac{d_2c_1V_*}{M^2} \int_\Omega |\nabla \widehat{V}|^2,
\end{aligned}$$

where $\widehat{U} = U - U_*$, $\widehat{V} = V - V_*$, and $M = \max\{a_1/b_1,\ a_2/c_2,\ \|U(x,0)\|_{L^\infty},\ \|V(x,0)\|_{L^\infty}\}$. Since $B > A > C$, it follows that $b_1b_2c_1c_2 > b_2^2c_1^2$, and we arrive at

$$\frac{d}{dt} E[U,V](t) \leq -\delta \left(\|\widehat{U}\|^2_{H_1(\Omega)} + \|\widehat{V}\|^2_{H_1(\Omega)} \right), \tag{4.12}$$

where

$$\delta = \min \left\{ \frac{b_1b_2c_1c_2 - b_2^2c_1^2}{2b_1b_2}, \frac{b_1b_2c_1c_2 - b_2^2c_1^2}{2c_1c_2}, \frac{d_1b_2U_*}{M^2}, \frac{d_2c_1V_*}{M^2} \right\}.$$

From (4.12) it is not hard to see that $(U(\cdot,t), V(\cdot,t)) \to (U_*, V_*)$ as $t \to \infty$, as (U_*, V_*) is the *unique global minimizer of* E. ☐

The situation for the *strong competition case*, i.e., $B < A < C$, is much more complicated. First, the constant (U_*, V_*) given by (4.11) is still a coexistence steady state of (4.10). However, it is now *unstable*, even in the ODE sense. Moreover, (4.10) can now have many nonconstant coexistence steady states; see, e.g., [GL]. In case both d_1, d_2 are large, it follows from Theorem 2.12 in Chapter 2 that (4.10) has no nonconstant steady state. Such a nonexistence fact actually holds as long as *one of the diffusion rates* d_1 *or* d_2 *is large*. This is proved in [LN1, Corollary 3.2]. The case when both d_1, d_2 are *not* large is not well understood, however. For instance, even *the existence or nonexistence of periodic solutions of* (4.10) *is open.*

4.4 Weak Competition in Heterogeneous Environment

Following Lou [L1], we consider the following Lotka–Volterra competition-diffusion system:

$$\begin{cases} U_t = d_1\Delta U + U(m_1(x) - U - cV) & \text{in } \Omega\times(0,\infty), \\ V_t = d_2\Delta V + V(m_2(x) - bU - V) & \text{in } \Omega\times(0,\infty), \\ \partial_\nu U = \partial_\nu V = 0 & \text{on } \partial\Omega\times(0,\infty), \end{cases} \tag{4.13}$$

where the spatial heterogeneity is reflected in $m_i(x) > 0$ on $\overline{\Omega}$ and $\not\equiv$ Constant, and we have normalized the intraspecific competition coefficients to be 1. Now, weak competition means

$$\frac{1}{b} > \frac{m_1(x)}{m_2(x)} > c \quad \text{for all} \quad x \in \overline{\Omega}. \tag{4.14}$$

In the previous section we proved that if $m_i \equiv$ Constant (i.e., in the spatially homogeneous case), there is a *unique* coexistence steady state (U_*, V_*) which is *globally asymptotically stable.* Now, the situation changes drastically. When the dispersal rates d_1, d_2 are both small, or both large, it still holds that (4.13) has a *unique* coexistence steady state (U_*, V_*) which is globally asymptotically stable. However, the proofs are much different. For the case that both d_1, d_2 are small, we refer readers to [HLM]. For the case that both d_1, d_2 are large, it may be proved by the same strategy as in [HLM]. Here, for the convenience of readers, we will include a proof due to Lou [L2]. The strategy is as follows. First, prove that the two semitrivial steady states $(\theta_{d_1,m_1}, 0)$ and $(0, \theta_{d_2,m_2})$ are unstable. Thus one can conclude from the theory of monotone flow that (4.13) has at least one coexistence steady state. Then show that *all* coexistence steady states are stable; thereby it follows that there is only one coexistence steady state.

Now we linearize the corresponding elliptic system of (4.13) at the semitrivial steady state $(\theta_{d_1,m_1}, 0)$:

$$\begin{cases} d_1\Delta\Psi_1 + \Psi_1(m_1(x) - 2\theta_{d_1,m_1}) - c\theta_{d_1,m_1}\Psi_2 + \lambda\Psi_1 = 0 & \text{in } \Omega, \\ d_2\Delta\Psi_2 + \Psi_2(m_2(x) - b\theta_{d_1,m_1}) \quad + \lambda\Psi_2 = 0 & \text{in } \Omega, \\ \partial_\nu\Psi_1 = \partial_\nu\Psi_2 = 0 & \text{on } \partial\Omega. \end{cases} \tag{4.15}$$

First, observe that all eigenvalues of (4.15) above are real-valued. For, if $\Psi_2 \not\equiv 0$, then λ is an eigenvalue of the second equation and thus is real-valued. Otherwise, $\Psi_2 \equiv 0$; then λ is an eigenvalue of the first equation, again real-valued.

Next, observe that for d_1 large

$$\int_\Omega \left(m_2(x) - b\theta_{d_1,m_1}\right) > 0$$

by (4.14) and the fact that $\int_\Omega \theta_{d_1,m_1} \to \int_\Omega m_1$ as $d_1 \to \infty$ (Theorem 4.1). Thus the first eigenvalue $\mu_1(d_2,\ m_2 - b\theta_{d_1,m_1})$ of the second equation in (4.15),

$$\begin{cases} d_2\Delta\widetilde{\Psi} + \widetilde{\Psi}(m_2(x) - b\theta_{d_1,m_1}) + \mu\widetilde{\Psi} = 0 & \text{in } \Omega, \\ \partial_\nu \widetilde{\Psi} = 0 & \text{on } \partial\Omega, \end{cases}$$

must be *negative* (Proposition 4.4). Now, choosing $\lambda = \mu_1(d_2,\ m_2 - b\theta_{d_1,m_1}) < 0$ and $\Psi_2 = \widetilde{\Psi}$ (the first eigenfunction corresponding to the eigenvalue $\mu_1(d_2, m_2 - b\theta_{d_1,m_1})$) in (4.15), we need to solve the first equation:

$$\begin{cases} d_1\Delta\Psi_1 + \Psi_1(m_1(x) - 2\theta_{d_1,m_1}) + \mu_1(d_2,\ m_2 - b\theta_{d_1,m_1})\Psi_1 = c\theta_{d_1,m_1}\widetilde{\Psi} & \text{in } \Omega, \\ \partial_\nu \Psi_1 = 0 & \text{on } \partial\Omega. \end{cases}$$

This obviously has a solution, as the first eigenvalue of the operator

$$d_1\Delta + (m_1 - 2\theta_{d_1,m_1})$$

has the property that $\mu_1(d_1,\ m_1 - 2\theta_{d_1,m_1}) > \mu_1(d_1,\ m - \theta_{d_1,m_1}) = 0$, by Proposition 4.4(iv), and therefore the operator

$$d_1\Delta + (m_1 - 2\theta_{d_1,m_1}) + \mu_1(d_2,\ m_2 - b\theta_{d_1,m_1})$$

is invertible since $\mu_1(d_2,\ m_2 - b\theta_{d_1,m_1})$ is negative. This shows that $\lambda = \mu_1(d_2,\ m_2 - b\theta_{d_1,m_1}) < 0$ is an eigenvalue of (4.15); consequently, $(\theta_{d_1,m_1}, 0)$ is *not* stable.

The instability of $(0, \theta_{d_2,m_2})$ can be handled similarly.

Now it follows from the theory of monotone flow that (4.13) has at least one coexistence steady state (U_*, V_*).

Next, we claim that, *for any coexistence steady state (U_*, V_*), we have, as $d_1, d_2 \to \infty$,*

$$(U_*, V_*) \to \left(\frac{\overline{m}_1 - c\overline{m}_2}{1 - bc},\ \frac{\overline{m}_2 - b\overline{m}_1}{1 - bc}\right). \tag{4.16}$$

It follows readily from the Maximum Principle that both U_*, V_* are uniformly bounded, independent of d_1, d_2. Passing to a subsequence if necessary, we have

$$(U_*, V_*) \to (\widetilde{U}, \widetilde{V})$$

in $C^2(\overline{\Omega})$ as $d_1, d_2 \to \infty$, where $\widetilde{U}$ and $\widetilde{V}$ are two nonnegative constants, by elliptic regularity estimates.

Integrating the equations for U_* and V_* and then passing to the limit, we obtain

$$\int_\Omega \widetilde{U}(m_1(x) - \widetilde{U} - c\widetilde{V}) = 0 = \int_\Omega \widetilde{V}(m_2(x) - b\widetilde{U} - \widetilde{V});$$

i.e.,

$$\begin{cases} \widetilde{U}(\overline{m}_1 - \widetilde{U} - c\widetilde{V}) = 0, \\ \widetilde{V}(\overline{m}_2 - b\widetilde{U} - \widetilde{V}) = 0. \end{cases} \tag{4.17}$$

Case 1. $\widetilde{U} = 0 = \widetilde{V}$.

This implies that U_* and $V_* \to 0$ in $C^2(\overline{\Omega})$, and consequently, for d_1, d_2 sufficiently large,

$$\int_\Omega U_*(m_1(x) - U_* - cV_*) > 0,$$

a contradiction.

Case 2. $\widetilde{U} = 0,\ \widetilde{V} > 0$.

From the second equation in (4.17) it follows that $\widetilde{V} = \overline{m}_2$. Setting $w = U^* / \|U^*\|_{L^\infty(\Omega)}$, we have $\|w\|_{L^\infty(\Omega)} = 1$ and

$$\Delta w = -\frac{1}{d_1} w(m_1(x) - U_* - cV_*)$$

with $\partial_\nu w = 0$ on $\partial\Omega$. As $d_1 \to \infty$, from elliptic regularity estimates we conclude that $w \to 1$ on $\overline{\Omega}$. Now

$$\begin{aligned} 0 &= \int_\Omega w(m_1(x) - U_* - cV_*) \\ &\to \int_\Omega (m_1(x) - c\widetilde{V}). \end{aligned}$$

Therefore, $\overline{m}_1 = c\widetilde{V} = c\overline{m}_2$, a contradiction.

Case 3. $\widetilde{U} > 0,\ \widetilde{V} = 0$.

This case may be handled in a similar fashion as Case 2.

Case 4. $\widetilde{U} > 0,\ \widetilde{V} > 0$.

Now it follows from (4.17) that

$$\widetilde{U} = \frac{\overline{m}_1 - c\overline{m}_2}{1 - bc}, \quad \widetilde{V} = \frac{\overline{m}_2 - b\overline{m}_1}{1 - bc},$$

and our assertion (4.16) is established.

It remains to show that *every coexistence steady state* (U_*, V_*) *is stable if both* d_1 *and* d_2 *are large.* Consider the linearized eigenvalue problem at (U_*, V_*):

$$\begin{cases} d_1\Delta\Psi_1 + \Psi_1(m_1(x) - 2U_* - cV_*) - cU_*\Psi_2 + \lambda\Psi_1 = 0 & \text{in } \Omega, \\ d_2\Delta\Psi_2 + \Psi_2(m_2(x) - bU_* - 2V_*) - bV_*\Psi_1 + \lambda\Psi_2 = 0 & \text{in } \Omega, \\ \partial_\nu\Psi_1 = \partial_\nu\Psi_2 = 0 & \text{on } \partial\Omega. \end{cases} \tag{4.18}$$

By the Krein–Rutman theory, (4.18) has a *real* principal eigenvalue λ_1, and all other eigenvalues λ have real parts $\geq \lambda_1$. Assume for contradiction that the principal eigenvalue $\lambda_1 \leq 0$.

Denote the corresponding eigenfunction by (Ψ_1, Ψ_2), normalized by $\|\Psi_1\|_{L^2}^2 + \|\Psi_2\|_{L^2}^2 = 1$. Then,

$$\begin{aligned}\lambda_1 &= d_1 \|\nabla \Psi_1\|_{L^2}^2 - \int_\Omega \Psi_1^2(m_1(x) - 2U_* - cV_*) + \int_\Omega cU_*\Psi_1\Psi_2 \\ &\quad + d_2\|\nabla\Psi_2\|_{L^2}^2 - \int_\Omega \Psi_2^2(m_2(x) - bU_* - 2V_*) + \int_\Omega bV_*\Psi_1\Psi_2 \\ &\geq -M,\end{aligned}$$

where M is a constant independent of d_1, d_2; i.e., $0 \geq \lambda_1 \geq -M$.

Now, letting $d_1, d_2 \to \infty$ in (4.18), by elliptic regularity estimates, we have, by passing to a subsequence if necessary,

$$(\Psi_1, \Psi_2) \to (\widetilde{\Psi}_1, \widetilde{\Psi}_2),$$

where $\widetilde{\Psi}_1$ and $\widetilde{\Psi}_2$ are two constants; thus $\|\widetilde{\Psi}_1\|_{L^2}^2 + \|\widetilde{\Psi}_2\|_{L^2}^2 = 1$. Now, integrating the two equations in (4.18) and then passing to limits, we obtain

$$\begin{cases}\widetilde{\Psi}_1(\overline{m}_1 - 2\widetilde{U} - c\widetilde{V}) - c\widetilde{U}\widetilde{\Psi}_2 + \lambda_1\widetilde{\Psi}_1 = 0, \\ \widetilde{\Psi}_2(\overline{m}_2 - b\widetilde{U} - 2\widetilde{V}) - b\widetilde{V}\widetilde{\Psi}_1 + \lambda_1\widetilde{\Psi}_2 = 0.\end{cases}$$

Simplifying, we have

$$\begin{pmatrix}\lambda_1 - \widetilde{U} & -c\widetilde{U} \\ -b\widetilde{V} & \lambda_1 - \widetilde{V}\end{pmatrix}\begin{pmatrix}\widetilde{\Psi}_1 \\ \widetilde{\Psi}_2\end{pmatrix} = \begin{pmatrix}0 \\ 0\end{pmatrix},$$

and therefore

$$\begin{vmatrix}\lambda_1 - \widetilde{U} & -c\widetilde{U} \\ -b\widetilde{V} & \lambda_1 - \widetilde{V}\end{vmatrix} = 0.$$

Thus $\lambda_1 > 0$ as $1 - bc > 0$, a contradiction, and our proof is complete. □

For d_1, d_2 neither large nor small, the situation changes drastically in heterogeneous environments. The first such phenomenon was discovered by Lou [L1]. Lou considered steady states of the special case $m_1 \equiv m_2$ in (4.13), namely, the elliptic system

$$\begin{cases}d_1\Delta U + U(m(x) - U - cV) = 0 & \text{in } \Omega, \\ d_2\Delta V + V(m(x) - bU - V) = 0 & \text{in } \Omega, \\ \partial_\nu U = \partial_\nu V = 0 & \text{on } \partial\Omega.\end{cases} \tag{4.19}$$

For simplicity, in the rest of this section we will write θ_{d_1} instead of $\theta_{d_1,m}$ and θ_{d_2} for $\theta_{d_2,m}$. Now the hypothesis (4.14) for weak competition becomes

$$\frac{1}{b} > 1 > c. \tag{4.20}$$

Lou [L1] has the following surprising result.

Theorem 4.7. *For every $b \in (b_*, 1)$ where*

$$b_* = \inf_{d>0} \frac{\overline{m}}{\overline{\theta_d}}, \tag{4.21}$$

there exists $c^* \in (0,1)$ small such that if $c \in (0, c^*)$, $(\theta_{d_1}, 0)$ is globally asymptotically stable for some $d_1 < d_2$.

In particular, for some $0 < b$, $c < 1$, and $d_1 < d_2$, U will wipe out V regardless of the initial conditions (as long as both $U(x,0)$ and $V(x,0)$ are nonnegative, nonzero), and coexistence is no longer possible even if the competition is weak! In fact, Lou's theorem gives a better result and more detailed information of the situation. (See [L1] for the details.) This theorem also shows the importance of the number $\overline{\theta}_d$, as promised at the beginning of this chapter. The number b_* is sharp, in some sense, as *if $b < b_*$, then $(\theta_{d_1}, 0)$ is always unstable for any d_1, d_2!*

The first step in establishing Lou's theorem is to prove *the local stability of $(\theta_{d_1}, 0)$ for some d_1, d_2 but for all $1 > b > b_*$ and $c < 1$.* Since $b > b_*$, there exist $0 < \underline{d} < \overline{d}$ such that

$$\int_\Omega (m(x) - b\theta_{d_1}) < 0$$

for *all* $d_1 \in (\underline{d}, \overline{d})$, and therefore by Theorem 4.2, $\lambda_1(m - b\theta_{d_1}) > 0$ for *all* $d_1 \in (\underline{d}, \overline{d})$, as it is obvious that $(m - b\theta_{d_1})(x_0) > 0$, where x_0 is a maximum point of θ_{d_1}.

Setting

$$\mathcal{S}_b = \left\{ (d_1, d_2) \,\middle|\, d_1 \in (\underline{d}, \overline{d}),\ d_2 > \frac{1}{\lambda_1(m - b\theta_{d_1})} \right\},$$

we claim that $(\theta_{d_1}, 0)$ is stable for all $(d_1, d_2) \in \mathcal{S}_b$ and $c \le 1$. To this end, we observe first from Theorem 4.2 that

$$\lambda_1(m - b\theta_{d_1}) < \lambda_1(m - \theta_{d_1}) = \frac{1}{d_1};$$

i.e., $d_1 < 1/\lambda_1(m - b\theta_{d_1})$. Hence, for $(d_1, d_2) \in \mathcal{S}_b$, we always have $d_2 > d_1$. Now we linearize (4.19) at $(\theta_{d_1}, 0)$:

$$\begin{cases} d_1 \Delta \Psi_1 + \Psi_1(m - 2\theta_{d_1}) - c\theta_{d_1}\Psi_2 + \lambda \Psi_1 = 0 & \text{in } \Omega, \\ d_2 \Delta \Psi_2 + \Psi_2(m - b\theta_{d_1}) + \lambda \Psi_2 = 0 & \text{in } \Omega, \\ \partial_\nu \Psi_1 = \partial_\nu \Psi_2 = 0 & \text{on } \partial\Omega. \end{cases} \tag{$\tilde{L}_1$}$$

Similarly as in Section 4.2, *all* eigenvalues λ of $(\tilde{L}_1)$ are real. Thus, if $\Psi_2 \not\equiv 0$, we then have

$$\lambda \ge \mu_1(d_2,\ m - b\theta_{d_1}) > 0$$

by Proposition 4.4(ii) and $1/d_2 < \lambda_1(m - b\theta_{d_1})$.

If $\Psi_2 \equiv 0$, then, from the first equation in $(\tilde{L}_1)$, it follows that

$$\lambda \ge \mu_1(d_1,\ m - 2\theta_{d_1}) > \mu_1(d_1,\ m - \theta_{d_1}) = 0$$

by Proposition (4.4)(iv). This finishes the proof of the local stability of $(\theta_{d_1}, 0)$ for *all* (d_1, d_2) in $\mathcal{S}_b$, $b > b_*$.

Next, the instability of $(0,\theta_{d_2})$ follows from standard arguments as in Section 4.2, and we will be brief here. Linearizing (4.19) at $(0,\theta_{d_2})$ we have

$$\begin{cases} d_1\Delta\Psi_1+\Psi_1(m(x)-c\theta_{d_2})+\lambda\Psi_1=0 & \text{in } \Omega, \\ d_2\Delta\Psi_2+\Psi_2(m(x)-2\theta_{d_2})-b\theta_{d_2}\Psi_1+\lambda\Psi_2=0 & \text{in } \Omega, \\ \partial_\nu\Psi_1=\partial_\nu\Psi_2=0 & \text{on } \partial\Omega. \end{cases} \tag{$\tilde{L}_2$}$$

We claim that $\lambda=\mu_1(d_1,m-c\theta_{d_2})$ *is a negative eigenvalue of* $(\tilde{L}_2)$.

First, observe that

$$\lambda=\mu_1(d_1,\, m-c\theta_{d_2})<\mu_1(d_2,\, m-c\theta_{d_2})<\mu_1(d_2,\, m-\theta_{d_2})=0,$$

again by Proposition 4.4. Letting Ψ_1 be the first eigenfunction of the first equation in $(\tilde{L}_2)$, we need to solve Ψ_2 from the second equation in $(\tilde{L}_2)$. But this follows readily from the fact that

$$\mu_1(d_2,m-2\theta_{d_2})>\mu_1(d_2,m-\theta_{d_2})=0$$

and that $\lambda=\mu_1(d_1,m-c\theta_{d_2})<0$. Thus $(\tilde{L}_2)$ has at least one negative eigenvalue $\lambda=\mu_1(d_1,m-c\theta_{d_2})$ and $(0,\theta_{d_2})$ is therefore unstable.

To finish the proof, we need to show that the system (4.19) does not have any coexistence steady state; then the global stability of $(\theta_{d_1},0)$ follows from the theory of monotone flow as before.

Lemma 4.8. *Fix* $b\in(b_*,1)$. *For any compact subset* $K\subseteq\mathcal{S}_b$, *there exists* $\delta>0$ *such that* (4.19) *has no coexistence steady state for* $(d_1,d_2)\in K$ *and* $0<c<\delta$.

Proof. Suppose for contradiction that such a $\delta>0$ does not exist. Then, there exists a sequence $(d_1^{(i)},d_2^{(i)})\in K$, $c_i\to 0$ such that the system (4.19) has a coexistence steady state (U_i,V_i), where $U_i>0$ and $V_i>0$. By standard elliptic regularity estimates, U_i,V_i are uniformly bounded in $C^{2,\alpha}(\overline{\Omega})$. Then, by passing to a subsequence if necessary, we may assume that $(d_1^{(i)},d_2^{(i)})\to(\widehat{d_1},\widehat{d_2})\in K$, $c_i\to 0$, and $U_i,V_i\to(\widehat{U},\widehat{V})$ in $C^{2,\alpha}(\overline{\Omega})$ with

$$\begin{cases} \widehat{d_1}\Delta\widehat{U}+\widehat{U}(m(x)-\widehat{U})=0 & \text{in } \Omega, \\ \widehat{d_2}\Delta\widehat{V}+\widehat{V}(m(x)-b\widehat{U}-\widehat{V})=0 & \text{in } \Omega, \\ \partial_\nu\widehat{U}=\partial_\nu\widehat{V}=0, & \text{on } \partial\Omega. \end{cases}$$

From the first equation, $\widehat{U}\equiv 0$ or $\widehat{U}\equiv\theta_{\widehat{d_1}}$, by Theorem 4.1.

Case 1: $\widehat{U}\equiv 0$.

This implies that $U_i\to 0$ in $C^{2,\alpha}(\overline{\Omega})$. Integrating the first equation of (4.19) for (U_i,V_i), we have

$$0\ge -d_1^{(i)}\int_\Omega\frac{|\nabla U_i|^2}{U_i^2}=\int_\Omega(m(x)-U_i-c_iV_i)\to\int_\Omega m,$$

a contradiction, since $m\ge 0$ and $\not\equiv 0$.

Case 2: $\widehat{U} \equiv \theta_{\widehat{d}_1}$.

Then $\widehat{V}$ satisfies

$$\begin{cases} \widehat{d}_2 \Delta \widehat{V} + \widehat{V}(m(x) - b\theta_{\widehat{d}_1} - \widehat{V}) = 0 & \text{in } \Omega, \\ \partial_\nu \widehat{V} = 0 & \text{on } \partial\Omega. \end{cases}$$

Since $(\widehat{d}_1, \widehat{d}_2) \in K$, $\widehat{d}_2 > 1/\lambda_1(m - b\theta_{\widehat{d}_1})$, which implies that $\widehat{V} \equiv 0$, i.e., $V_i \to 0$ in $C^{2,\alpha}(\overline{\Omega})$. Setting $\widetilde{V}_i = V_i / \|V_i\|_{L^\infty(\Omega)}$ and sending $i \to \infty$, we have

$$\widehat{d}_2 \Delta \widetilde{V} + \widetilde{V}(m - b\theta_{\widehat{d}_1}) = 0$$

(by passing to a subsequence if necessary), where $\widetilde{V} > 0$ since $\|\widetilde{V}\|_{L^\infty(\Omega)} = 1$. This guarantees that $\widehat{d}_2 = 1/\lambda_1(m - b\theta_{\widehat{d}_1})$, a contradiction, and our proof is complete. □

Comparing Lou's theorem to Theorem 4.5, we see that Lou's theorem, while not containing the phenomenon in Theorem 4.5, still seems general, as it allows more flexibilities. An unfortunate fact is that it requires c to be small, while in Theorem 4.5, $c = 1$. In a recent paper [LmN2], the following result is obtained.

Theorem 4.9. *For any $\varepsilon > 0$, there exists a $\delta = \delta(\varepsilon) > 0$ such that for $b \in (1-\delta, 1)$, $c \in [0,1]$, $\varepsilon < d_1 < \frac{1}{\varepsilon}$, and $d_2 > d_1 + \varepsilon$, $(\theta_{d_1}, 0)$ is globally asymptotically stable.*

In particular, when $\epsilon \to 0$ in Theorem 4.9, the phenomenon "Slower diffuser always prevails!" (Theorem 4.5) is a limiting case.

It also seems interesting to note that the competition coefficient c in the result above could be bigger than b!

Lam and Ni [LmN2] also has the following result, complementing Theorem 4.9.

Theorem 4.10. *For any $b, c \in (0,1)$, there exists an $\varepsilon > 0$ such that if $|d_1 - d_2| < \varepsilon$, then there exists a unique positive (coexistence) steady state (U_*, V_*) of* (4.19)*, where $U_* > 0$ and $V_* > 0$. Moreover, (U_*, V_*) is globally asymptotically stable; and, as $d_1, d_2 \to d > 0$,*

$$(U_*, V_*) \to \frac{1}{1 - bc}(1 - c, 1 - b)\theta_d.$$

Theorems 4.7, 4.9, and 4.10 together seem to illustrate the important and delicate role of the diffusion. (See Figure 4.2.)

The proof of Theorem 4.10 consists of two steps. First, for the special case $d_1 = d_2 = d$, we establish the existence and uniqueness of a coexistence steady state (U_*, V_*), $U_* > 0$, $V_* > 0$, and the instability of the two semitrivial steady states $(\theta_{d_1}, 0)$ and $(0, \theta_{d_2})$. Then, Theorem 4.10, in the special case, follows from the theory of monotone flow.

Next, we show the local stability of the coexistence steady state (U_*, V_*) when $d_1 = d_2 = d$, which implies, by the Implicit Function theorem, the existence of a coexistence steady state for $d_1 \neq d_2$, but both are close to d. The general case of Theorem 4.10 then follows from the following lemma, whose proof is omitted here.

Lemma 4.11. *For any $b, c \in (0,1)$, there exist two small positive numbers ε, r such that* (4.19) *has no coexistence steady states in $B_r(S)$ in $H_1(\Omega)$, where $S = (0,0)$, $(\theta_{d_1}, 0)$, or $(0, \theta_{d_2})$, if $|d_1 - d_2| < \varepsilon$.*

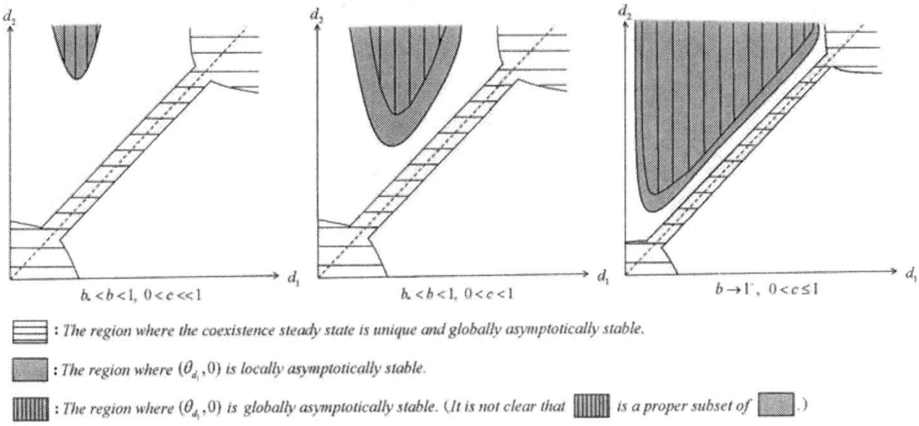

Figure 4.2.

In proving Theorem 4.9 we need to show that, for the parameters b, c, d_1, d_2 in the ranges specified, (4.19) has no coexistence steady states. To this end, a result similar to Lemma 4.11 is obtained. In addition, a shadow system is treated since now d_2 could tend to ∞.

To conclude this chapter, we consider the following special case:

$$\begin{cases} U_t = d_1\Delta U + U(m(x) - U - bV) & \text{in } \Omega \times (0,\infty), \\ V_t = d_2\Delta V + V(m(x) - bU - V) & \text{in } \Omega \times (0,\infty), \\ \partial_\nu U = \partial_\nu V = 0 & \text{on } \partial\Omega \times (0,\infty), \end{cases}$$

where $b = 1 - \delta$, $\delta > 0$ small. Here, U and V are two "equal" competitors, in some sense, except for their dispersal rates. Then Theorem 4.9 above of Lam and Ni [LmN2] indicates that, with $d_2 > 0$ fixed, the species U does not seem to fare better as d_1 decreases from d_2 to 0, especially when d_1 is very small. In fact, from the arguments presented in this section, it is not hard to see that $(\theta_{d_1}, 0)$ is indeed unstable for d_1 sufficiently small!

Chapter 5

Beyond Diffusion: Directed Movements, Taxis, and Cross-Diffusion

The logistic equation and the Lotka–Volterra competition system have been around for a long time, and adding diffusion terms seems only the first step while considering spatial effects. In reality, few species disperse completely randomly, and it seems reasonable to assume that individuals could detect and respond to some of the environmental cues (e.g., changes in resources or population pressure) and thereby exhibit various directed/biased movements or taxis.

A basic equation in population dynamics is

$$u_t = \nabla \cdot [d\nabla u \pm u\nabla\psi(E(x,t))] + f(x,u),$$

where u represents the population density, $d > 0$ is the dispersal rate, ψ is increasing, and E reflects environmental influences that could also depend on u. Note that the first term here is diffusion, while the second term represents the "directed movements" or "taxis." Examples for ψ include

$$\psi(E) = kE, \quad k\log E, \quad \text{or} \quad \frac{kE^m}{1+aE^m},$$

where k, a, m are positive constants. When the positive sign in front of the term $u\nabla\psi(E(x,t))$ is used, we refer to the movement as "negative taxis." When the negative sign is adopted, we have a "positive taxis."

In this chapter, we will discuss several examples in this direction. In the first example, we continue to explore the "Slower diffuser always prevails!" phenomenon studied in Chapter 4, but in a different direction; namely, we shall assume that one of the two competing species has some "intelligence" now—it can move up along the gradient direction of the resource term $m(x)$. This was first proposed by Belgacem and Cosner [BC] for single equations and then by Cantrell, Cosner, and Lou for competition systems [CCL1].

Our second example again deals with the classical Lotka–Volterra competition systems, but now in a *homogeneous* environment. Following Shigesada, Kawasaki, and Teramoto [SKT], we consider a cross-diffusion system modeling segregation phenomena of two competing species, which takes into account the interspecific population pressures—an example of "negative taxis."

The third example is the well-known Keller–Segel chemotaxis system modeling the aggregation stage in the reproduction of cellular slime molds when the environment is poor in resources. This example illustrates the effects of "positive taxis."

In the three examples above, only the first one involves spatial inhomogeneity *explicitly*, while in the second and third examples, in addition to the (random) diffusion, convection terms appear naturally, but the systems remain autonomous. A common feature in all these examples is that they all lack variational structures; that is, none of them is the Euler–Lagrange equation of a variational functional. In many ways, this makes them more interesting and challenging.

5.1 Directed Movements in Population Dynamics

When the environment is spatially inhomogeneous, we have seen the interesting phenomenon that the slower diffuser always prevails in the competition of two otherwise identical species with random dispersals. In the last chapter, to better understand this phenomenon, we studied the interaction between the dispersal rates and the competition abilities in spatially inhomogeneous environments. In this section, we return to this phenomenon but pursue in a different direction.

Since few species disperse completely randomly, in this section, we shall assume, between the two competing species U and V, that the species U has the ability of moving up along the gradient of the resource term $m(x)$, while for comparison purposes V is still assumed to disperse randomly. Therefore, U and V satisfy the following system—first proposed by Cantrell, Cosner, and Lou [CCL1] based on an earlier single equation model by Belgacem and Cosner [BC]:

$$\begin{cases} U_t = \nabla\cdot(d_1\nabla U - \alpha U\nabla m) + U(m(x) - U - V) & \text{in } \Omega\times(0,T), \\ V_t = d_2\Delta V + V(m(x) - U - V) & \text{in } \Omega\times(0,T), \\ d_1\partial_\nu U - \alpha U\partial_\nu m = 0 = \partial_\nu V & \text{on } \partial\Omega\times(0,T), \end{cases} \tag{5.1}$$

where $T \le \infty$, $m(x) \not\equiv$ Constant is again assumed to be nonnegative for simplicity, and $\alpha \ge 0$ measures the strength of the "directed" movement of U. Many fundamental properties of (5.1) have been obtained in the two important papers [CCL1, CCL2]. In this section we will describe recent progress on this interesting system.

5.1.1 Single Equations

The system (5.1) again has two semitrivial steady states $(\theta_{d_1,\alpha}, 0)$ and $(0, \theta_{d_2})$, where $\theta_{d_1,\alpha}$ is the *unique positive* solution of

$$\begin{cases} \nabla\cdot(d_1\nabla u - \alpha u\nabla m) + u(m(x) - u) = 0 & \text{in } \Omega, \\ d_1\partial_\nu u - \alpha u\partial_\nu m = 0 & \text{on } \partial\Omega, \end{cases} \tag{5.2}$$

and θ_{d_2} is defined as in Theorem 4.1 (with the subindex m suppressed). The fact that (5.2) has a *unique positive stable* solution, for every $d_1 > 0, \alpha \ge 0$, is proved in [BC], [CL].

Furthermore, under a suitable hypothesis on $m(x)$, the solution $\theta_{d_1,\alpha}$ exhibits spike-layer concentration phenomena as $\alpha \to \infty$.

Again, to facilitate our needs for the competition system (5.1), we need to consider a slightly more general single equation than (5.2), namely,

$$\begin{cases} u_t = \nabla\cdot(d\nabla u - \alpha u\nabla h) + u(h(x)-u) & \text{in } \Omega\times(0,T), \\ d\partial_\nu u - \alpha u\partial_\nu h = 0 & \text{on } \partial\Omega\times(0,T), \end{cases} \tag{5.3}$$

where $\alpha \geq 0$ and h is *positive somewhere* in Ω; i.e., $\{x\in\Omega\,|\,h(x)>0\}$ is nonempty. Then Belgacem and Cosner [BC] showed the following.

Theorem 5.1. *Suppose that $h(x)$ is positive somewhere in Ω. Then, for every α sufficiently large,* (5.3) *has a unique positive steady state $\theta_{d,\alpha}$ which is globally asymptotically stable among nonnegative nonzero solutions.*

Thus the existence, uniqueness, and global stability of the positive steady state of (5.3) hold for quite general h.

In [CCL2], Cantrell, Cosner, and Lou showed that *if the set of all critical points of h has Lebesgue measure* 0*, then the unique steady state $\theta_{d,\alpha}\to 0$ in $L^2(\Omega)$ as $\alpha\to\infty$*. Based on this fact, the following conjecture was proposed in [CCL2].

Conjecture. *The concentration set of $\theta_{d,\alpha}$, as $\alpha\to\infty$, is precisely the set of all positive local maximum points of $h(x)$.*

This conjecture was recently established under an additional hypothesis on h by Lam [Lm1]:

(H1) $h\in C^3(\overline{\Omega})$, $\partial_\nu h\leq 0$ on $\partial\Omega$, and h is positive somewhere in Ω.

(H2) h has only finitely many local maximum points on $\overline{\Omega}$, all being strict local maxima and located in the interior of Ω.

(H3) $\Delta h(x_0)>0$ if $x_0\in\overline{\Omega}$ is a local minimum or a saddle point of h.

Letting $\mathcal{H}$ be the set of *all positive strict* local maximum points of h in $\overline{\Omega}$, Lam [Lm1] showed the following.

Theorem 5.2. *Let $\theta_{d,\alpha}$ be the unique positive steady state of* (5.3).

(i) *For each $x_0\in\mathcal{H}$ and any ball $B_r(x_0)$ centered at x_0 with radius r,*

$$\liminf_{\alpha\to\infty}\sup_{B_r(x_0)}\theta_{d,\alpha}\geq h(x_0).$$

(ii) *Assume in addition that* (H1), (H2), *and* (H3) *hold. Then, for any compact subset $K\subseteq\overline{\Omega}\setminus\mathcal{H}$, there exists a constant $\gamma=\gamma(K)$ such that for all $x\in K$*

$$\theta_{d,\alpha}(x)\leq e^{-\gamma\alpha}.$$

In particular, $\theta_{d,\alpha}\to 0$ uniformly and exponentially in K as $\alpha\to\infty$.

The profile of $\theta_{d,\alpha}$ can be determined for α large as well. In the case $n=1$, this was obtained by Lam and Ni [LmN1], inspired by the work of Chen and Lou [CnL] on the system.

Theorem 5.3. *Let $\Omega=(-1,1)$. Suppose that* (H1) *holds and that all critical points of h are nondegenerate. Then, for any $r>0$ small and $x_0\in\mathcal{H}$, we have, as $\alpha\to\infty$,*

(i) $\theta_{d,\alpha}\to 0$ *uniformly and exponentially in* $\Omega\setminus\bigcup_{x\in\mathcal{H}}(x_0-r,x_0+r)$;

(ii) $\left\|\theta_{d,\alpha}-\sqrt{2}h(x_0)e^{\alpha[h(x)-h(x_0)]/d}\right\|_{L^\infty(x_0-r,x_0+r)}\to 0.$

The proof of Theorem 5.3 is lengthy but elementary. The starting point is the following formula obtained by integrating the equation in (5.3):

$$\frac{\theta_{d,\alpha}(x)}{\theta_{d,\alpha}(x_\alpha)}=\exp\left[\frac{\alpha}{d}(h(x)-h(x_\alpha))-\frac{1}{d}\int_{x_\alpha}^{x}\frac{1}{\theta_{d,\alpha}(z)}\left(\int_{-1}^{z}\theta_{d,\alpha}(h-\theta_{d,\alpha})\right)dz\right] \tag{5.4}$$

for any $x,x_\alpha\in(-1,1)$. Thus the profile of $\theta_{d,\alpha}$ is determined if the integral in (5.4) is small. This follows from a careful estimate of the length of the set

$$\{x\in(-1,1)\,|\,\theta_{d,\alpha}(x)>\alpha^{-k}\}$$

(i.e., where $\theta_{d,\alpha}$ is not *too* small) for various exponents $k>0$.

The multidimensional version of Theorem 5.3 was recently obtained by Lam [Lm2] using a different method.

Theorem 5.4 (see [Lm2]). *Suppose that $\mathcal{H}\subseteq\Omega$ and that* (H1) *and* (H3) *hold. Assume further that all critical points of h are nondegenerate. Then, as $\alpha\to\infty$, for any fixed $r>0$,*

(i) $\theta_{d,\alpha}\to 0$ *uniformly and exponentially on any compact subset* $K\subseteq\overline{\Omega}\setminus\mathcal{H}$;

(ii) $\|\theta_{d,\alpha}-\sqrt{2^n}h(x_0)e^{\alpha[h(x)-h(x_0)]/d}\|_{L^\infty(B_r(x_0))}\to 0.$

Lam's proof consists of two main ingredients: an L^∞-estimate of $\theta_{d,\alpha}$ for all α large (i.e., independent of α large), and a Liouville-type theorem for the limiting equation near $x_0\in\mathcal{H}$.

For the uniform L^∞-estimate of $\theta_{d,\alpha}$, we first integrate the equation in (5.3) to obtain

$$\int_\Omega\theta_{d,\alpha}^2=\int_\Omega h\,\theta_{d,\alpha}. \tag{5.5}$$

Then applying the Harnack inequality to (5.3) in the neighborhood $B_{R/\sqrt{\alpha}}(x_0)$, $x_0\in\mathcal{H}$, and estimating $\theta_{d,\alpha}$ carefully in the three subsets $B_{R/\sqrt{\alpha}}(x_0)$, $B_r(x_0)\setminus B_{R/\sqrt{\alpha}}(x_0)$, and $\Omega\setminus B_r(x_0)$ separately, we get from (5.5) that

$$\left(\sup_\Omega\theta_{d,\alpha}\right)^2\le C_1\sup_\Omega\theta_{d,\alpha}+C_2,$$

where the constants C_1, C_2 are independent of α large. Now the uniform bound for $\theta_{d,\alpha}$ follows immediately.

Next, near each $x_0 \in \mathcal{H}$, rescale the variables in (5.3) as follows:

$$x = x_0 + y\sqrt{\frac{d}{\alpha}} \quad \text{and} \quad w = e^{\alpha[h(x)-h(x_0)]}u;$$

then w satisfies

$$\nabla \cdot \left(de^{\alpha[h(x)-h(x_0)]/d}\nabla w\right) + u(h-u)\frac{d}{\alpha} = 0,$$

which converges to

$$\nabla \cdot \left(e^{\frac{1}{2}y^T D^2 h(x_0) y}\nabla w\right) = 0, \quad y \in \mathbb{R}^n, \tag{5.6}$$

as $\alpha \to \infty$, where $D^2 h(x_0)$ is the Hessian matrix of h at $x_0 \in \mathcal{H}$ and y^T is the transpose of y.

Lam's Liouville-type theorem for (5.6) reads as follows.

Theorem 5.5. *Every nonnegative weak solution $w \in W^{1,2}_{loc}(\mathbb{R}^n)$ of (5.6) is a constant.*

In general, some kind of asymptotic behavior is needed for such Liouville-type results to hold; see, e.g., [BCN2]. We refer interested readers to [Lm2] for a detailed proof of Theorem 5.5.

5.1.2 Systems

When $d_1 < d_2$ and α is small, U still wipes out V in (5.1); i.e., $(\theta_{d_1,\alpha}, 0)$ is *globally asymptotically stable for $U(x,0) \geq 0, \not\equiv 0$—slower diffuser still prevails!* [CCL1].

When $d_1 = d_2$, the situation becomes delicate, as the system (5.1) becomes *degenerate* when $d_1 = d_2$ and $\alpha = 0$; that is, the steady states of (5.1) now form a one-parameter family $\{(\beta\theta_d, (1-\beta)\theta_d) \mid 0 \leq \beta \leq 1\}$, where $d = d_1 = d_2$. (The proof is left as an exercise for interested readers.) Very interesting and delicate results in this direction have been obtained by [CCL1, CCL2].

Here we shall focus on the case $\alpha > 0$ large. Again the first set of interesting results is due to [CCL2]. In particular, they showed the following.

Theorem 5.6. *Suppose that (in addition to the hypothesis that $m(x) \geq 0$, $\not\equiv$ Constant)*

(M1) *the set of all critical points of m has measure 0;*

(M2) *there exists $x_0 \in \overline{\Omega}$ such that $m(x_0) = \max_{\overline{\Omega}} m$ is a strict local maximum.*

Then, for any $d_1, d_2 > 0$, (5.1) has a stable coexistence steady state (U_α, V_α), $U_\alpha > 0$, $V_\alpha > 0$, for α large.

In fact, Cantrell, Cosner, and Lou proved much more in [CCL2]—they showed that under (M1) and (M2), *the semitrivial steady states $(\theta_{d_1,\alpha}, 0)$ and $(0, \theta_{d_2})$ are unstable for α large, and one always has coexistence for α large.* Furthermore, *for any coexistence steady state (U_α, V_α), it always holds that*

$$U_\alpha \to 0 \text{ in } L^2(\Omega) \quad \text{and} \quad V_\alpha \rightharpoonup \theta_{d_2} \text{ in } H_2(\Omega)$$

as $\alpha \to \infty$. Based on this result, they proposed the following.

Conjecture. *Under* (M1) *and* (M2), (5.1) *has a unique coexistence steady state* (U_α, V_α) *which is globally asymptotically stable, and, as* $\alpha \to \infty$, U_α *concentrates at all local maximum points of* $m(x)$ *in* $\overline{\Omega}$.

For the rest of this section, we will always assume that (M1) and (M2) hold. In [CnL] important progress was made when m has a unique critical point in $\overline{\Omega}$ which is a nondegenerate global maximum point.

Theorem 5.7 (see [CnL]). *Suppose that* m *has a unique critical point* x_0 *in* Ω *which is a nondegenerate global maximum point of* m *on* $\overline{\Omega}$ *and that* $\partial_\nu m \le 0$ *on* $\partial\Omega$. *Let* (U_α, V_α) *be a coexistence steady state of* (5.1). *Then, as* $\alpha \to \infty$,

(i) $V_\alpha \to \theta_{d_2}$ *in* $C^{1,\beta}(\overline{\Omega})$;

(ii) $\left\| U_\alpha(x) \exp\left[-\frac{\alpha}{2d_1}(x-x_0)^T D^2 m(x_0)(x-x_0) \right] - 2^{\frac{n}{2}} \left[m(x_0) - \theta_{d_2}(x_0) \right] \right\|_{L^\infty(\Omega)} \to 0.$

The factor $2^{\frac{n}{2}}[m(x_0) - \theta_{d_2}(x_0)]$ comes from the integral constraint

$$\int_\Omega U_\alpha (m - V_\alpha - U_\alpha) = 0$$

and the profile of U_α

$$U_\alpha(x) \sim \exp\left[\frac{\alpha}{2d_1}(x-x_0)^T D^2 m(x_0)(x-x_0) \right]$$

near x_0.

For general $m(x)$, in the one-dimensional case $n = 1$, using the result Theorem 5.3 for single equations, the profile of U_α in the Conjecture was obtained by Lam and Ni [LmN1]. It is convenient to let $\mathfrak{M}$ be the set of all local maximum points of m on $\overline{\Omega}$.

Theorem 5.8. *Suppose that* $\Omega = (-1,1)$, $\partial_\nu m \le 0$ *on* $\partial\Omega$, *and* $\mathfrak{M} \subseteq \Omega$ *and that all local maximum points of* m *in* $\mathfrak{M}$ *are nondegenerate. Let* (U_α, V_α) *be a coexistence steady state of* (5.1). *Then, as* $\alpha \to \infty$,

(i) $V_\alpha \to \theta_{d_2}$ *in* $C^{1,\beta}(\overline{\Omega})$;

(ii) *for any* $x_0 \in \mathfrak{M}$ *and any* $r > 0$ *small*

$$\left\| U_\alpha(x) - \max\left\{ \sqrt{2}(m - \theta_{d_2})(x_0), 0 \right\} \exp\left[\frac{\alpha}{2d_1} m''(x_0)(x-x_0)^2 \right] \right\|_{L^\infty(B_r(x_0))} \to 0;$$

(iii) *for any compact subset* $K \subseteq \overline{\Omega} \setminus \mathfrak{M}$, $U_\alpha \to 0$ *uniformly and exponentially on* K.

Theorem 5.8 shows that the conjecture above needs to modify slightly, as U_α will *not* survive at those local maximum points of m where $m \le \theta_{d_2}$! In other words, local maximum points of m could serve as "traps" for U_α if θ_{d_2} is greater than or equal to m there! (And, it is easy to see that sometimes there are such points.) (See Figure 5.1.)

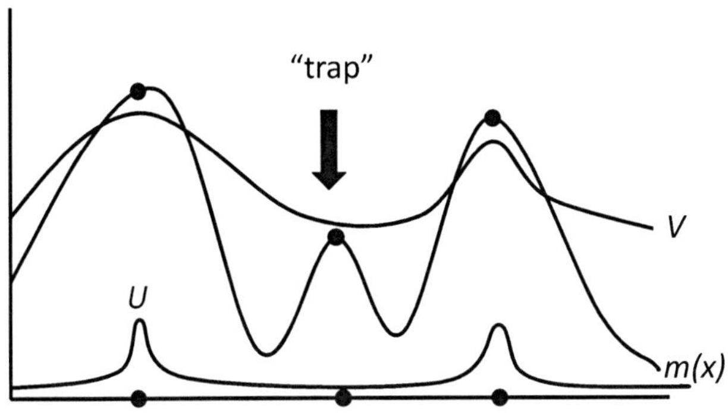

Figure 5.1.

It is interesting to observe that for α large, V_α behaves as if U_α didn't exist—indeed, $U_\alpha \to 0$ in L^p as $\alpha \to \infty$.

The proof of Theorem 5.8 follows that of Theorem 5.3 and is different from that of Theorem 5.7 in [CnL].

Likewise, using Theorem 5.4, Lam [Lm2] succeeded in extending Theorem 5.8 to the multidimensional case under a slightly stronger hypothesis.

Theorem 5.9. *Suppose that $\partial_\nu m \le 0$ on $\partial\Omega$, $\mathfrak{M} \subseteq \Omega$, and all critical points of m are nondegenerate. Furthermore, assume that $\Delta m(x_0) > 0$ for every saddle point x_0 of m. Let (U_α, V_α) be a coexistence steady state of (5.1). Then, as $\alpha \to \infty$,*

(i) *$V_\alpha \to \theta_{d_2}$ in $C^{1,\beta}(\overline{\Omega})$;*

(ii) *for any $x_0 \in \mathfrak{M}$ and any $r > 0$ small*

$$\left\| U_\alpha(x) - \max\left\{2^{\frac{n}{2}}(m-\theta_{d_2})(x_0), 0\right\} \exp\left[\frac{\alpha}{2d_1}(x-x_0)^T D^2 m(x_0)(x-x_0)\right]\right\|_{L^\infty(B_r(x_0))} \to 0;$$

(iii) *for any compact subset $K \subseteq \overline{\Omega}\setminus\mathfrak{M}$, $U_\alpha \to 0$ on K uniformly and exponentially.*

The condition (M2) can be removed; see [CnHL, Lemmas 6.4 and 6.5]. The case when (M1) does not hold is very interesting and, in fact, seems important and worth pursuing.

5.1.3 Other Related Models

A more recent model developed by [CCL3] proposes a mathematical formulation of the concept "ideal free distribution." More precisely, in a one-species single equation case, it takes the following form:

$$\begin{cases} u_t = \nabla\cdot[d\nabla u - \alpha u \nabla(m-u)] + u(m-u) & \text{in } \Omega\times(0,T), \\ d\partial_\nu u - \alpha u \partial_\nu(m-u) = 0 & \text{on } \partial\Omega\times(0,T), \end{cases} \tag{IFD}$$

where again the resource term $m(x)$ is *nonconstant*, and $T \leq \infty$. The basic idea here is that the species u can sense the magnitude of the "available" resources $m(x) - u$ and move up its gradient. In [CCL3], among other things, the following remarkable result is obtained: *Suppose that* $m \in C^{2,\beta}(\overline{\Omega})$ *and* $m > 0$ *in* $\overline{\Omega}$. *Then, for every* α *large,* (IFD) *has a unique positive solution* u_α, *which is globally asymptotically stable, and* $u_\alpha \to m$ *in* $C^2(\overline{\Omega})$ *as* $\alpha \to \infty$.

In other words, as $\alpha \to \infty$, the species u *tends to distribute itself to match perfectly the resources* $m(x)$ *on* $\overline{\Omega}$.

It seems natural to ask how such a species would fare under competition. This amounts to solving a cross-diffusion system, and it seems that so far little is known in this direction.

5.2 Cross-Diffusions in Lotka–Volterra Competition System

For autonomous single equations, diffusion is generally regarded as a trivializing process. This is no longer the case for autonomous 2×2 reaction-diffusion systems, as we have seen in the activator-inhibitor system mentioned in Chapter 3, in which the *gap* of the diffusion rates forces the system to create strikingly nontrivial spike-layer solutions. However, for this to happen, *reaction terms* are essential as well. In Section 4.3, we have seen that the classical Lotka–Volterra competition system

$$\begin{cases} U_t = d_1 \Delta U + U(a_1 - b_1 U - c_1 V) & \text{in } \Omega \times (0,\infty), \\ V_t = d_2 \Delta V + V(a_2 - b_2 U - c_2 V) & \text{in } \Omega \times (0,\infty), \\ \partial_\nu U = \partial_\nu V = 0 & \text{on } \partial\Omega \times (0,\infty), \end{cases} \tag{5.7}$$

where all constants a_i, b_i, c_i, and d_i, $i = 1,2$, are positive and U, V are nonnegative, *possesses no nonconstant steady state, no matter what* d_1, d_2 *are, in the weak competition case*, i.e., if

$$B > A > C, \tag{5.8}$$

where $A = a_1/a_2$, $B = b_1/b_2$, $C = c_1/c_2$.

It seems not entirely reasonable, *even in the homogeneous environment*, to assume that individuals *only* move around randomly. In particular, it seems reasonable to take into consideration the population pressures created by the competitors while modeling segregation phenomena for two competing species. In this direction, Shigesada, Kawasaki, and Teramoto [SKT] proposed in 1979 the following model:

$$\begin{cases} U_t = \Delta[(d_1 + \rho_{11} U + \rho_{12} V)U] + U(a_1 - b_1 U - c_1 V) & \text{in } \Omega \times (0,T), \\ V_t = \Delta[(d_2 + \rho_{21} U + \rho_{22} V)V] + V(a_2 - b_2 U - c_2 V) & \text{in } \Omega \times (0,T), \\ \partial_\nu U = \partial_\nu V = 0 & \text{on } \partial\Omega \times (0,T), \end{cases} \tag{5.9}$$

where ρ_{12} and ρ_{21} represent the interspecific population pressures, ρ_{11} and ρ_{22} denote the intraspecific population pressures, and all are assumed to be nonnegative. Here in this chapter, for simplicity we shall focus mostly on the case that the "self-diffusions" $\rho_{11} =$

$\rho_{22} = 0$, i.e., the cross-diffusion system

$$\begin{cases} U_t = \Delta[(d_1 + \rho_{12}V)U] + U(a_1 - b_1U - c_1V) & \text{in } \Omega \times (0,T), \\ V_t = \Delta[(d_2 + \rho_{21}U)V] + V(a_2 - b_2U - c_2V) & \text{in } \Omega \times (0,T), \\ \partial_\nu U = \partial_\nu V = 0 & \text{on } \partial\Omega \times (0,T). \end{cases} \tag{5.10}$$

It is worth noting that even local existence (in time) for (5.9) or (5.10) is highly nontrivial and was resolved in a series of long papers of Amann [A1, A2] around 1990.

Considerable work has been done on the global existence (in time) of (5.9) under various hypotheses. However, almost all of them require that *at least one of the self-diffusion rates* ρ_{11}, ρ_{22} *be positive* (in order to "control" the cross-diffusion coefficients ρ_{12} and ρ_{21}) and therefore do not apply to the cross-diffusion system (5.10). To this date, only three small instances of progress have been reported in the literature concerning the global existence of (5.10). First, Kim [Km] in 1984 stated that in $n = 1$, if $d_1 = d_2$, then solutions of (5.10) (with nonnegative initial values $U(x,0)$ and $V(x,0)$) exist for all $t > 0$. (Those solutions were proved to be bounded by Shim [Sh] in 2002.) Second, Deuring [D] in 1987 showed global existence for solutions of (5.10) with small initial values. Finally, in 1998, for $n \leq 2$, it was proved by [LNW] that *if* $\rho_{21} = 0$, *then all solutions of* (5.10) *exist globally in time* $t > 0$. The global existence of (5.10) remains an outstanding open problem 30 years after the system was first proposed.

We now turn to the effect of cross-diffusion on steady states. To illustrate the significance of cross-diffusions, we again go to the weak competition case (i.e., $B > A > C$) since in this case (5.10) has no nonconstant steady states if both $\rho_{12} = \rho_{21} = 0$. (We refer readers to [Hu] for some interesting discussions on the ecological significance of coexistence, "competition-exclusion," and weak/strong competitions. One point of view is that whether "competition-exclusion" holds in nature is a matter of interpretation. See [Wm].) Recent work of Lou and Ni [LN1, LN2] shows that, indeed, *if one of the two cross-diffusion rates, say* ρ_{12}, *is large, then* (5.10) *will have nonconstant steady states, provided that* d_2 *belongs to a proper range.* On the other hand, *if both* ρ_{12} *and* ρ_{21} *are small, then* (5.10) *will have no nonconstant steady states under the condition* (5.8). This shows that cross-diffusion does seem to help create patterns.

In the "strong competition" case, i.e., $B < A < C$, even the situation of steady state solutions of (5.7) becomes more interesting and complicated and is *not* completely understood. Nonetheless, cross-diffusion still has similar effects in helping to create more nontrivial patterns of (5.10). We refer interested readers to [LN1, LN2] for details.

So far in this section, we have only touched upon the existence and nonexistence of nonconstant steady states. It seems natural and important to ask if we can derive any qualitative properties (such as the spike layers in Chapter 3) of those steady states. Our first step in this direction is to classify *all the possible (limiting) steady states* as one of the cross-diffusion pressures tends to infinity.

Theorem 5.10 (see [LN2]). *Suppose for simplicity that* $\rho_{21} = 0$. *Suppose further that* $B \neq A \neq C$, $n \leq 3$, *and* $\frac{a_2}{d_2} \neq \lambda_k$ *for all* k, *where* λ_k *is the* k*th eigenvalue of* Δ *on* Ω *with zero Neumann boundary data. Let* (U_j, V_j) *be a nonconstant steady state solution of* (5.10) *with* $\rho_{12} = \rho_{12,j}$. *Then, by passing to a subsequence if necessary, either* (i) *or* (ii) *holds as* $\rho_{12,j} \to \infty$, *where*

(i) $(U_j, \frac{\rho_{12,j}}{d_1}V_j) \to (U,V)$ *uniformly,* $U>0$, $V>0$, *and*

$$\begin{cases} d_1\Delta[(1+V)U]+U(a_1-b_1U)=0 & \text{in } \Omega, \\ d_2\Delta V+V(a_2-b_2U)=0 & \text{in } \Omega, \\ \partial_\nu U=\partial_\nu V=0 & \text{on } \partial\Omega; \end{cases} \tag{5.11}$$

(ii) $(U_j, V_j) \to (\frac{\zeta}{w}, w)$ *uniformly,* ζ *is a positive constant,* $w>0$, *and*

$$\begin{cases} d_2\Delta w+w(a_2-c_2w)-b_2\zeta=0 & \text{in } \Omega, \\ \partial_\nu w=0 & \text{on } \partial\Omega, \\ \displaystyle\int_\Omega \frac{1}{w}\left(a_1-\frac{b_1\zeta}{w}-c_1w\right)=0. \end{cases} \tag{5.12}$$

The proof is quite lengthy. The most important step in the proof is to obtain a priori bounds on steady states of (5.10) that are independent of ρ_{12}.

We ought to remark that both systems (5.11) and (5.12) possesses spike-layer solutions. For instance, using a suitable change of variables, the equation in (5.12) may be transformed into (3.1) with $\varepsilon^2=d_2$ and $p=2$. Thus our results in Section 3.1 apply. Perhaps we ought to point out that, in fact, what is important is the ratio of cross-diffusion versus diffusion ρ_{12}/d_1 in which d_1 can also vary. A deeper classification result is obtained in [LN2] as $\rho_{12}\to\infty$ in (5.10) in terms of various possibilities of ρ_{12}/d_1 and d_1. To see how (3.1) turns up in (5.12), at least heuristically, we proceed as follows. Formally, setting

$$\zeta=U_*V_* \quad \text{and} \quad w=V_*-\varphi, \tag{5.13}$$

where (U_*, V_*) is the constant coexistence steady state given by (4.11), we have

$$d_2\Delta\varphi-(c_2V_*-b_2U_*)\varphi+c_2\varphi^2=0. \tag{5.14}$$

Rescaling (5.14) we obtain (3.1), provided that

$$c_2V_*-b_2U_*>0,$$

which is equivalent to

$$\begin{cases} \frac{1}{2}(B+C)>A & \text{if } B>A>C, \\ \frac{1}{2}(B+C)<A & \text{if } B<A<C. \end{cases}$$

Note that in (5.13) we need $w>0$, or $V_*>\varphi$. In $n=1$ this is guaranteed by

$$A>\frac{1}{4}(B+3C).$$

Under these conditions, our results in Section 3.1 imply that (5.14) has spike-layer solutions for d_2 small. Observe that those solutions tend to 0 as $d_2\to 0$ except at isolated points. Let φ be, e.g., the solution of (3.1) guaranteed by Theorem 3.3. Then the pair (w, U_*V_*) satisfies the differential equation with the homogeneous Neumann boundary condition in (5.12), and it *almost* satisfies the integral constraint in (5.12) since w is close to V_* a.e. for

d_2 small. It is then not hard to find a solution, for d_2 small, near the pair $(w, U_* V_*)$ by the Implicit Function theorem, as was done in [LN2].

Although (5.11) is still an elliptic system, it is a bit easier to analyze than the original one. We refer interested readers to [LN2, Section 5] for details.

It turns out that both alternatives (i) and (ii) in Theorem 5.10 occur under suitable conditions. Therefore, to understand the steady states of (5.10) a good model would be (5.11) or (5.12), at least when ρ_{12} is large. In the work of Lou, Ni, and Yotsutani [LNY], we were able to achieve an almost *complete* understanding of the "shadow" system (5.12) for $n = 1$ (and Ω is an interval, say, $(0,1)$). To illustrate our results, we include the following theorems.

Theorem 5.11. *Suppose $B < C$. Then (5.12) does not have any nonconstant solution if either one of the following two conditions holds:*

(i) $d_2 \geq a_2/\pi^2$;

(ii) $A \leq B$.

Theorem 5.12. *Suppose $B < C$. Then (5.12) has a nonconstant solution if $d_2 < a_2/\pi^2$ and $A \geq (B+C)/2$.*

The case $d_2 < a_2/\pi^2$ and $B < A < (B+C)/2$ is more delicate—existence holds for d_2 closer to a_2/π^2, while nonexistence holds when d_2 is near 0.

The behavior of solutions is also obtained for d_2 close to one of the two endpoints, 0 or a_2/π^2.

Theorem 5.13.

(i) *As $d_2 \to a_2/\pi^2$, $(w, \zeta) \to (0,0)$ in such a way that*

$$\frac{\zeta}{w} \to \frac{a_2(1+\mu)}{2[\mu + (1-\mu)\sin^2(\pi x/2)]}$$

uniformly on $[0,1]$, where $\mu = (2A/B) - 1 - 2\sqrt{(A/B)^2 - (A/B)} \in (0,1]$.

(ii) *As $d_2 \to 0$, we have*

(α) *if $A < \frac{B+3C}{4}$, then*

$$\zeta \to \frac{a_2^2(B-A)(A-C)}{b_2 c_2 (B-C)^2},$$

$$w(0) \to 2\frac{a_2}{c_2}\frac{A-(B+3C)/4}{B-C},$$

$$w \to \frac{a_2}{c_2}\frac{B-A}{B-C} \quad \text{on } (0,1];$$

(β) *if $A \geq \frac{B+3C}{4}$, then $\zeta \to \frac{3}{16}\frac{a_2^2}{b_2 c_2}$, $w(0) \to 0$, and $w \to \frac{3a_2}{4c_2}$ on $(0,1]$.*

It seems interesting to note that the limits in (β) above are independent of a_1, b_1, c_1.

Our method of proof here is a bit unusual: we convert the problem of solving (w,ζ) of (5.12) to a problem of solving its "representation" in a different parameter space. This is done first *without* the integral constraint in (5.12). Then we use the integral constraint to find the "solution curve" in the new parameter space as the diffusion rate d_2 varies. This method turns out to be very powerful, as it gives fairly precise information about the solution.

Of course, our ultimate goal is to be able to obtain the steady states of (5.10) from our knowledge of the simpler limiting system (5.11) or (5.12). This turns out to be possible, at least in the one-dimensional case $\Omega = (0,1)$, as the next two results show. (For simplicity, we shall assume that $\rho_{21} = 0$ in the next two theorems.)

Theorem 5.14 (see [LN2]). *Suppose that $A > B$. There exists a small $d^* > 0$ such that for any $d_2 \in (0,d^*)$, we can find a large $\tilde{d} > 0$ such that if $d_1 \geq \tilde{d}$ is fixed, then there exists a large $\alpha > 0$ such that if $\rho_{12} > \alpha$,* (5.10) *has a nonconstant positive steady state (U,V), with $(U,\rho_{12}V) \to (\bar{U},\bar{V})$ uniformly in $[0,1]$ as $\rho_{12} \to \infty$, where $(\bar{U},\bar{V})$ is a nonconstant positive solution of* (5.11).

Theorem 5.15 (see [LN2]). *Suppose that $d_1 > 0$ is fixed and that either $A \in (\frac{1}{2}(B+C),(\frac{1}{4}B+\frac{3}{4}C))$ or $A \in ((\frac{1}{4}B+\frac{3}{4}C),\frac{1}{2}(B+C))$. There exists a small $d^* > 0$ such that for $d_2 \in (0,d^*)$, we can find a large $\alpha > 0$ such that if $\rho_{12} > \alpha$,* (5.10) *has a nonconstant positive steady state (U,V) with $(U,V) \to (\frac{\zeta}{w},w)$ as $\rho_{12} \to \infty$, where $w > 0$ and nonconstant and (w,ζ) is a solution of* (5.12).

The proofs of Theorems 5.14 and 5.15 involve careful analysis of the linearized systems of (5.11) and (5.12) at their nonconstant positive solutions.

Very recently, great progress was made by Yotsutani [Y]—the delicate remaining cases are resolved and a complete understanding of (5.12) for $n = 1$ is achieved.

Finally, we mention that stability analysis of the steady states obtained in this section is generally quite delicate and difficult. Excellent progress in this direction has been made by Wu [Wu].

5.3 A Chemotaxis System

Chemotaxis is the oriented movement of cells in response to chemicals in their environment. Cellular slime molds (amoebae) are one such example—they release a certain chemical, c-AMP, move toward its higher concentration, and eventually form aggregates. Letting $U(x,t)$ be the population of amoebae at place x and at time t, and $V(x,t)$ be the concentration of this chemical, Keller and Segel [KS] proposed the following model to describe the chemotactic aggregation stage of amoebae:

$$\begin{cases} U_t = d_1 \Delta U - \chi \nabla \cdot \left[U \nabla \psi(V) \right] & \text{in } \Omega \times (0,T), \\ V_t = d_2 \Delta V - aV + bU & \text{in } \Omega \times (0,T), \\ \partial_\nu U = 0 = \partial_\nu V & \text{on } \partial\Omega \times (0,T), \\ U(x,0) = U_0(x), \quad V(x,0) = V_0(x) & \text{in } \Omega, \end{cases} \tag{5.15}$$

where the constants χ, a, and b are positive. Examples for the "sensitivity function" ψ include $\psi(V) = kV$, $k \log V$, and $kV^2/(1+V^2)$, where $k > 0$ is a constant. There have been many other models proposed by various authors over the years; we shall mention only (5.15), as our purpose here is just to introduce systems in this direction to interested readers.

A large amount of work has focused on the linear case $\psi(V) = kV$, and much is known in this case, at least for the low spatial dimensions, $n = 1$ or 2. The mathematical phenomena exhibited here are rich, from nontrivial steady states to blow-up dynamics.

For the logarithmic case $\psi(V) = k \log V$, Nagai and Senba [NS] recently proved global existence for a modified parabolic-elliptic system in case $n = 2$. Observe that in (5.15) the total population is always conserved; that is, for all $t > 0$ we have

$$\int_\Omega U(x,t)dx \equiv \int_\Omega U_0(x)dx.$$

Therefore to study the steady states of (5.15) for the case $\psi(V) = \log V$ we consider the following elliptic system:

$$\begin{cases} d_1 \Delta U - \chi \nabla \cdot (U \nabla \log V) = 0 & \text{in } \Omega, \\ d_2 \Delta V - aV + bU = 0 & \text{in } \Omega, \\ \partial_\nu U = 0 = \partial_\nu V & \text{on } \partial\Omega, \\ \frac{1}{|\Omega|} \int_\Omega U(x)dx = \bar{U} & \text{(prescribed).} \end{cases} \tag{5.16}$$

With $p = \chi/d_1$, it is not hard to show that $U = \lambda V^p$ for some constant $\lambda > 0$. Thus, setting $\epsilon^2 = d_2/a$, $\mu = (b\lambda/a)^{1/(p-1)}$, and $w = \mu V$, we see that w satisfies (3.1), i.e.,

$$\begin{cases} \epsilon^2 \Delta w - w + w^p = 0 & \text{in } \Omega, \\ \partial_\nu w = 0 & \text{on } \partial\Omega, \end{cases} \tag{5.17}$$

and our previous results for (3.1) apply. To obtain a solution pair for (5.16) from a solution of (5.17), simply set

$$U = \frac{\bar{U}|\Omega| w^p}{\int_\Omega w^p} \quad \text{and} \quad V = \frac{\bar{V}|\Omega| w}{\int_\Omega w}$$

with $\bar{V} = b\bar{U}/a$. In this way, *we obtain spike-layer steady states for the chemotaxis system* (5.16) *when* d_2/a *is small and* $1 < \chi/d_1 < \frac{n+2}{n-2}$ *(*$= \infty$ *if* $n = 1, 2$*).* Although many believe the particular steady state corresponding to the "least-energy" solution of (3.1) is stable, its proof has thus far eluded us.

Bibliography

[A1] H. Amann, *Dynamic theory of quasilinear parabolic equations.* II. *Reaction-diffusion systems*, Differential Integral Equations **3** (1990), 13-75.

[A2] H. Amann, *Nonhomogeneous linear quasilinear elliptic and parabolic boundary value problems*, Function Spaces, Differential Operators and Nonlinear Analysis, H. Schmeisser and H. Triebel, eds., Teubner Texte Math. **133**, Teubner, Stuttgart (1993), 9-126.

[AMN1] A. Ambrosetti, A. Malchiodi, and W.-M. Ni, *Solutions, concentrating on spheres, to symmetric singularly perturbed problems*, C. R. Math. Acad. Sci. Paris **335** (2002), 145-150.

[AMN2] A. Ambrosetti, A. Malchiodi, and W.-M. Ni, *Singularly perturbed elliptic equations with symmetry: Existence of solutions concentrating on spheres*, I, Comm. Math. Phys. **235** (2003), 427-466.

[AMN3] A. Ambrosetti, A. Malchiodi, and W.-M. Ni, *Singularly perturbed elliptic equations with symmetry: Existence of solutions concentrating on spheres*, II, Indiana Univ. Math. J. **53** (2004), 297-329.

[AR] A. Ambrosetti and P. Rabinowitz, *Dual variational methods in critical point theory and applications*, J. Funct. Anal. **14** (1973), 349-381.

[AmC] L. Ambrosio and X. Cabré, *Entire solutions of semilinear elliptic equations in* $\mathbb{R}^3$ *and a conjecture of De Giorgi*, J. Amer. Math. Soc. **13** (2000), 725-739.

[BB] R. Banuelos and K. Burdzy, *On the "hot spots" conjecture of J. Rauch*, J. Funct. Anal. **164** (1999), 1-33.

[BC] F. Belgacem and C. Cosner, *The effects of dispersal along environmental gradients on the dynamics of populations in heterogeneous environments*, Canad. Appl. Math. Q. **3** (1995), 379-397.

[BCN1] H. Berestycki, L. A. Caffarelli, and L. Nirenberg, *Symmetry for elliptic equations in a half space*, Boundary Value Problems for Partial Differential Equations and Applications, C. Baiocchi, ed., RMA Res. Notes Appl. Math. **29**, Masson, Paris (1993), 27-42.

[BCN2] H. Berestycki, L. A. Caffarelli, and L. Nirenberg, *Further qualitative properties for elliptic equations in unbounded domains*, Ann. Scuola Norm. Sup. Pisa Cl. Sci. (4) **25** (1997), 69-94.

[BN] H. Berestycki and L. Nirenberg, *On the method of moving planes and the sliding method*, Bol. Soc. Brasil. Mat. (N.S.) **22** (1991), 1-37.

[B] T. Bhattacharya, *Radial symmetry of the first eigenfunction for the p-Laplacian in the ball*, Proc. Amer. Math. Soc. **104** (1988), 169-174.

[Bi] L. Bieberbach, $\Delta u = e^u$ *und die automorphen Functionen*, Math. Ann. **77** (1916), 173-212.

[BL] H. J. Brascamp and E. Lieb, *On extensions of Brun-Minkowski and Prekoph-Leindler theorems, including inequalities for log concave functions and with an application to the diffusion equation*, J. Funct. Anal. **22** (1976), 366-389.

[BrN] H. Brézis and L. Nirenberg, *Positive solutions of nonlinear elliptic equations involving critical Sobolev exponents,* Comm. Pure Appl. Math. **36** (1983), 437-477.

[BS] J. Busca and B. Sirakov, *Symmetry results for semilinear elliptic systems in the whole space*, J. Differential Equations **163** (2000), 41-56.

[CF] L. A. Caffarelli and A. Friedman, *Convexity of solutions of semilinear elliptic equations*, Duke Math. J. **52** (1985), 431-456.

[CGS] L. A. Caffarelli, B. Gidas, and J. Spruck, *Asymptotic symmetry and local behavior of semilinear elliptic equations with critical Sobolev growth*, Comm. Pure Appl. Math. **42** (1989), 271-297.

[CK] T. K. Callahan and E. Knobloch, *Pattern formation in three-dimensional reaction-diffusion systems*, Phys. D **132** (1999), 339-362.

[CC] R. S. Cantrell and C. Cosner, Spatial Ecology via Reaction-Diffusion Equations, Wiley Series in Mathematical and Computational Biology, John Wiley & Sons, New York, 2003.

[CCL1] R. S. Cantrell, C. Cosner, and Y. Lou, *Movement toward better environments and the evolution of rapid diffusion*, Math. Biosci. **204** (2006), 199-214.

[CCL2] R. S. Cantrell, C. Cosner, and Y. Lou, *Advection-mediated coexistence of competing species*, Proc. Roy. Soc. Edinburgh Sect. A **137** (2007), 497-518.

[CCL3] R. S. Cantrell, C. Cosner, and Y. Lou, *Approximating the ideal free distribution via reaction-diffusion-advection equations*, J. Differential Equations **245** (2008), 3687-3703.

[CaH] R. G. Casten and C. J. Holland, *Instability results for a reaction diffusion equation with Neumann boundary conditions*, J. Differential Equations **27** (1978), 266-273.

[CDBD] V. Castets, E. Dulos, J. Boissonade, and P. De Kepper, *Experimental evidence of a sustained standing Turing-type nonequilibrium chemical pattern*, Phys. Rev. Lett. **64** (1990), 2953-2956.

[Ch] S. Chandrasekhar, An Introduction to the Study of Stellar Structure, University of Chicago Press, Chicago, IL, 1939.

[CLi] W. Chen and C. Li, *Qualitative properties of solutions to some nonlinear elliptic equations in* $\mathbb{R}^n$, Duke Math. J. **71** (1993), 427-439.

[CnHL] X. Chen, R. Hambrock, and Y. Lou, *Evolution of conditional dispersal: A reaction-diffusion-advection model*, J. Math. Biol. **57** (2008), 361-386.

[CnL] X. Chen and Y. Lou, *Principal eigenvalue and eigenfunctions of an elliptic operator with large advection and its application to a competition model*, Indiana Univ. Math. J. **57** (2008), 627-658.

[CN] K.-S. Cheng and W.-M. Ni, *On the structure of the conformal Gaussian curvature equation on* $\mathbb{R}^2$, Duke Math. J. **62** (1991), 721-737.

[Co] C. V. Coffman, *A nonlinear boundary value problem with many positive solutions*, J. Differential Equations **54** (1984), 429-437.

[CE] P. Collet and J.-P. Eckmann, *Space-time behaviour in problems of hydrodynamic type: A case study*, Nonlinearity **5** (1992), 1265-1302.

[CL] C. Cosner and Y. Lou, *Does movement toward better environments always benefit a population?*, J. Math. Anal. Appl. **277** (2003), 489-503.

[CH] R. Courant and D. Hilbert, Methods of Mathematical Physics, Vol. I, Interscience, New York, 1953.

[DPR] L. Damascelli, F. Pacella, and M. Ramaswamy, *Symmetry of ground states of p-Laplace equations via the moving plane method*, Arch. Ration. Mech. Anal. **148** (1999), 291-308.

[DG] E. N. Dancer and Z.-M. Guo, *Some remarks on the stability of sign changing solutions*, Tohoku Math. J. (2) **47** (1995), 119-225.

[DH] M. Dauge and B. Helffer, *Eigenvalues variation,* I *and* II, J. Differential Equations **104** (1993), 243-297.

[DKW] M. del Pino, M. Kowalczyk, and J. Wei, *On De Giorgi conjecture in dimensions* $N \geq 9$, Ann. Math, to appear.

[D] P. Deuring, *An initial-boundary value problem for a certain density-dependent diffusion system*, Math. Z. **194** (1987), 375-396.

[DN] W.-Y. Ding and W.-M. Ni, *On the existence of positive entire solutions of a semilinear elliptic equation*, Arch. Rational Mech. Anal. **91** (1986), 283-308.

[DHMP] J. Dockery, V. Hutson, K. Mischaikow, and M. Pernarowski, *The evolution of slow dispersal rates: A reaction diffusion model*, J. Math. Biol. **37** (1998), 61-83.

[DGK] A. Doelman, R. A. Gardner and T. J. Kaper, *A stability index analysis of* 1*-D patterns of the Gray-Scott model*, Mem. Amer. Math. Soc. **155** (2002), no. 737.

[E] S. D. Eidel'man, Parabolic Systems, North–Holland, Amsterdam, 1969.

[F] P. C. Fife, Dynamics of Internal Layers and Diffusive Interfaces, CBMS-NSF Regional Conf. Ser. in Appl. Math. **53**, SIAM, Philadelphia, 1988.

[FL] B. Franchi and E. Lanconelli, *Radial symmetry of the ground states for a class of quasilinear elliptic equations*, Nonlinear Diffusion Equations and Their Equilibrium States, Vol. I, W.-M. Ni, L.A. Peletier, and J. Serrin, eds., Math. Sci. Res. Inst. Publ. **12**, Springer-Verlag, New York (1988), 287-292.

[GG1] N. Ghoussoub and C. Gui, *On a conjecture of De Giorgi and some related problems.* Math. Ann. **311** (1998), 481-491.

[GG2] N. Ghoussoub and C. Gui, *On De Giorgi's conjecture in dimensions* 4 *and* 5, Ann. of Math. (2) **157** (2003), 313-334.

[GNN1] B. Gidas, W.-M. Ni, and L. Nirenberg, *Symmetry and related properties via the Maximum Principle*, Comm. Math. Phys. **68** (1979), 209-243.

[GNN2] B. Gidas, W.-M. Ni, and L. Nirenberg, *Symmetry of positive solutions of nonlinear elliptic equations in* $\mathbb{R}^n$, Mathematical Analysis and Applications, Part A, Adv. in Math. Suppl. Stud. **7a**, Academic Press, New York (1981), 369-402.

[GM] A. Gierer and H. Meinhardt, *A theory of biological pattern formation*, Kybernetik **12** (1972), 30-39.

[GS] P. Gray and S. K. Scott, *Autocatalytic reactions in the isothermal, continuous stirred tank reactor: Oscillations and instabilities in the system* $A + 2B \to 3B, B \to C$, Chem. Engrg. Sci. **39** (1984), 1087-1097.

[Gd] P. Grisvard, Elliptic Problems in Nonsmooth Domains, Monogr. Stud. Math. **24**, Pitman (Advanced Publishing Program), Boston, MA, 1985.

[Gr] M. Grossi, *Uniqueness of the least-energy solution for a semilinear Neumann problem*, Proc. Amer. Math. Soc. **128** (1999), 1665-1672.

[GPW] M. Grossi, A. Pistoia, and J. Wei, *Existence of multipeak solutions for a semilinear Neumann problem via nonsmooth critical point theory*, Calc. Var. Partial Differential Equations **11** (2000), 143-175.

[G] C. Gui, *Multi-peak solutions for a semilinear Neumann problem*, Duke Math. J. **84** (1996), 739-769.

[GL] C. Gui and Y. Lou, *Uniqueness and nonuniqueness of coexistence states in the Lotka-Volterra competition model*, Comm. Pure Appl. Math. **47** (1994), 1571-1594.

[GNW1] C. Gui, W.-M. Ni, and X. Wang, *On the stability and instability of positive steady states of a semilinear heat equation*, Comm. Pure Appl. Math. **45** (1992), 1153-1181.

[GNW2] C. Gui, W.-M. Ni, and X. Wang, *Further study on a nonlinear heat equation*, J. Differential Equations **169** (2001), 588-613.

[GW1] C. Gui and J. Wei, *Multiple interior peak solutions for some singularly perturbed problems*, J. Differential Equations **158** (1999), 1-27.

[GW2] C. Gui and J. Wei, *On multiple mixed interior and boundary peak solutions for some singularly perturbed Neumann problems*, Canad. J. Math. **52** (2000), 522-538.

[GWW] C. Gui, J. Wei and M. Winter, *Multiple boundary peak solutions for some singularly perturbed Neumann problems*, Ann. Inst. H. Poincaré Anal. Non Linéaire **17** (2000), 47-82.

[HR1] J. K. Hale and G. Raugel, *Reaction-diffusion equation on thin domains*, J. Math. Pures Appl. (9) **71** (1992), 33-95.

[HR2] J. K. Hale and G. Raugel, *A damped hyperbolic equation on thin domains*, Trans. Amer. Math. Soc. **329** (1992), 185-219.

[HS] J. K. Hale and K. Sakamoto, *Shadow systems and attractors in reaction-diffusion equations*, Appl. Anal. **32** (1989), 287-303.

[HV] J. K. Hale and J. M. Vegas, *A nonlinear parabolic equation with varying domain*, Arch. Rational Mech. Anal. **86** (1984), 99-123.

[He] P. Hess, Periodic-Parabolic Boundary Value Problems and Positivity, Pitman Res. Notes Math. Ser. **247**, Longman, Harlow, 1991.

[H] M. Hirsch, *Stability and convergence in strongly monotone dynamical systems*, J. Reine Angew. Math. **383** (1988), 1-53.

[Ho] D. Horstmann, *From* 1970 *until present: The Keller-Segel model in chemotaxis and its consequences,* Jahresber. Deutsch. Math.-Verein. **105** (2003), 103-165.

[Hs] S. B. Hsu, *Limiting behavior for competing species*, SIAM J. Appl. Math. **34** (1978), 760-763.

[HY] B. Hu and H. M. Yin, *Semilinear parabolic equations with prescribed energy*, Rend. Circ. Mat. Palermo (2) **44** (1995), 479-505.

[Hu] G. E. Hutchinson, An Introduction to Population Theory, Yale University Press, New Haven, CT, 1978.

[HLM] V. Hutson, Y. Lou, and K. Mischaikow, *Convergence in competition models with small diffusion coefficients*, J. Differential Equations **211** (2005), 135-161.

[J] J. Jang, *Symmetry of positive solutions to semilinear elliptic problems in half space*, Nonlinear Anal. **49** (2002), 613-621.

[JNT] J. Jang, W.-M. Ni, and M. Tang, *Global bifurcation and structure of Turing patterns in 1-D Lengyel-Epstein model*, J. Dynam. Differential Equations **16** (2004), 297-320.

[Jg] H. Jiang, *Global existence of solutions of an activator-inhibitor system*, Discrete Contin. Dyn. Syst. A **14** (2006), 737-751.

[JM1] S. Jimbo and Y. Morita, *Remarks on the behavior of certain eigenvalues on a singularly perturbed domain with several thin channels*, Comm. Partial Differential Equations **17** (1992), 523-552.

[JM2] S. Jimbo and Y. Morita, *Stability of nonconstant steady-state solutions to a Ginzburg-Landau equation in higher space dimensions*, Nonlinear Anal. **22** (1994), 753-770.

[JS] S. L. Judd and M. Silber, *Simple and superlattice Turing patterns in reaction-diffusion systems: Bifurcation, bistability, and parameter collapse*, Phys. D **136** (2000), 45-65.

[KN] Y. Kabeya and W.-M. Ni, *Stationary Keller-Segel model with linear sensitivity*, Publ. Res. Inst. Math. Sci. **1025** (1998), 44-65.

[K] S. Kamin, *On stabilisation of solutions of the Cauchy problem for parabolic equations*, Proc. Roy. Soc. Edinburgh Sect. A **76** (1976/77), 43-53.

[Ke] J. Keener, *Activators and inhibitors in pattern formation*, Stud. Appl. Math. **59** (1978), 1-23.

[KS] E. F. Keller and L. A. Segel, *Initiation of slime mold aggregation viewed as an instability*, J. Theoret. Biol. **26** (1970), 399-415.

[Kn] A. Kennington, *Power concavity and boundary value preblems*, Indiana Univ. Math. J. **34** (1985), 688-704.

[Km] J. U. Kim, *Smooth solutions to a quasilinear system of diffusion equations for a certain population model*, Nonlinear Anal. **8** (1984), 1121-1144.

[KW] K. Kishimoto and H. F. Weinberger, *The spatial homogeneity of stable equilibria of some reaction-diffusion systems on convex domains*, J. Differential Equations **58** (1985), 15–21.

[Ko] N. Korevaar, *Convex solutions to nonlinear elliptic and parabolic boundary value problems*, Indiana Univ. Math. J. **32** (1983), 603-614.

[KoL] N. Korevaar and J. Lewis, *Convex solutions of certain elliptic equations have constant rank Hessians*, Arch. Rational Mech. Anal. **97** (1987), 19-32.

[Lm1] K.-Y. Lam, *Concentrating phenomena of a semilinear elliptic equation with large advection in an ecological model*, J. Differential Equations **250** (2011), 161-181.

[Lm2] K.-Y. Lam, *Limiting profiles of semilinear elliptic equations with large advection in population dynamics* II, preprint (2011).

[LmN1] K.-Y. Lam and W.-M. Ni, *Limiting profiles of semilinear elliptic equations with large advection in population dynamics*, Discrete Contin. Dyn. Syst. A **28** (2010), 1051-1067.

[LmN2] K.-Y. Lam and W.-M. Ni, in preparation.

[LMPS] K. J. Lee, W. D. McCormick, J. E. Pearson, and H. L. Swinney, *Experimental observation of self-replicating spots in a reaction-diffusion system*, Nature **369** (1994), 215-218.

[LE1] I. Lengyel and I. R. Epstein, *Modeling of Turing structures in the chlorite-iodide-malonic-acid-starch reaction system*, Science **251** (1991), 650-652.

[LE2] I. Lengyel and I. R. Epstein, *A chemical approach to designing Turing patterns in reaction-diffusion systems*, Proc. Nat. Acad. Sci. U.S.A. **89** (1992), 3977-3979.

[LNa] F. Li and K. Nakashima, *Transition layers for a spatially inhomogeneous Allen-Cahn equation in multi-dimensional domains*, preprint (2011).

[LNN] F. Li, K. Nakashima, and W.-M.Ni, *Stability from the point of view of diffusion, relaxation and spatial inhomogeneity*, Discrete Contin. Dyn. Syst. A **20** (2008), 259-274.

[LfN] F. Li and W.-M. Ni, *On the global existence and finite time blow-up of shadow systems*, J. Differential Equations **247** (2009), 1762-1776.

[LCQ] M. Li, S. Chen, and Y. Qin, *Boundedness and blow-up for the general activator-inhibitor model*, Acta Math. Appl. Sinica (English Ser.) **11** (1995), 59-68.

[Ly] Yanyan Li, *Existence of many positive solutions of semilinear elliptic equations on annulus*, J. Differential Equations **83** (1990), 348-367.

[Li] Yi Li, *Asymptotic behavior of positive solutions of equation* $\Delta u + K(x)u^p = 0$ *in* $\mathbb{R}^n$, J. Differential Equations **95** (1992), 304-330.

[LiN] Yi Li and W.-M. Ni, *Radial symmetry of positive solutions of nonlinear elliptic equations in* $\mathbb{R}^n$, Comm. Partial Differential Equations **18** (1993), 1043-1054.

[LnN] C.-S. Lin and W.-M. Ni, *Stability of solutions of semilinear diffusion equations*, preprint (1986).

[LNT] C.-S. Lin, W.-M. Ni, and I. Takagi, *Large amplitude stationary solutions to a chemotaxis system*, J. Differential Equations **72** (1988), 1-27.

[LT] C.-S. Lin and I. Takagi, *Method of rotating planes applied to a singularly perturbed Neumann problem*, Calc. Var. Partial Differential Equations **13** (2001), 519-536.

[LW] C.-S. Lin and J. Wei, *Uniqueness of multiple-spike solutions via the method of moving planes*, Pure Appl. Math. Q. **3** (2007), 689-735.

[L1] Y. Lou, *On the effects of migration and spatial heterogeneity on single and multiple species*, J. Differential Equations **223** (2006), 400-426.

[L2] Y. Lou, private communication (2009).

[L3] Y. Lou, *Some challenging mathematical problems in evolution of dispersal and population dynamics*, Tutorials in Mathematical Biosciences. **IV**, 171-205, Lecture Notes in Math. **1922**, Springer, Berlin, 2008.

[LN1] Y. Lou and W.-M. Ni, *Diffusion, self-diffusion and cross-diffusion*, J. Differential Equations **131** (1996), 79-131.

[LN2] Y. Lou and W.-M. Ni, *Diffusion vs. cross-diffusion: An elliptic approach*, J. Differential Equations **154** (1999), 157-190.

[LNW] Y. Lou, W.-M. Ni, and Y. Wu, *On the global existence of a cross-diffusion system*, Discrete Contin. Dynam. Systems A **4** (1998), 193-203.

[LNY] Y. Lou, W.-M. Ni, and S. Yotsutani, *On a limiting system in the Lotka-Volterra competition with cross-diffusion*, Discrete Contin. Dyn. Syst. A **10** (2004), 435-458.

[MX] X. Ma and L. Xu, *The convexity of solution of a class Hessian equation in bounded convex domain in* $\mathbb{R}^3$, J. Funct. Anal. **255** (2008), 1713-1723.

[Mg] K. Maginu, *Stability of stationary solutions of a semilinear parabolic partial differential equation*, J. Math. Anal. Appl. **63** (1978), 224-243.

[MM] F. Mahmondi and A. Malchiodi, *Concentration on minimal submanifolds for a singularly perturbed Neumann problem*, Adv. Math. **209** (2007), 460-525.

[M] A. Malchiodi, *Solutions concentrating at curves for some singularly perturbed elliptic problems*, C. R. Math. Acad. Sci. Paris **338** (2004), 775-780.

[MM1] A. Malchiodi and M. Montenegro, *Boundary concentration phenomena for a singularly perturbed elliptic problem*, Comm. Pure Appl. Math. **55** (2002), 1507-1568.

[MM2] A. Malchiodi and M. Montenegro, *Multidimensional boundary-layers for a singularly perturbed Neumann problem*, Duke Math. J. **124** (2004), 105-143.

[MNW] A. Malchiodi, W.-M. Ni, and J. Wei, *Multiple clustered layer solutions for semilinear Neumann problems*, Ann. Inst. H. Poincaré Anal. Non Linéaire **22** (2005), 143-163.

[Ma1] H. Matano, *Convergence of solutions of one-dimensional semilinear parabolic equations*, J. Math. Kyoto Univ. **18** (1978), 221-227.

[Ma2] H. Matano, *Asymptotic behavior and stability of solutions of semilinear diffusion equations*, Publ. Res. Inst. Math. Sci. **15** (1979), 410-454.

[My] Y. Miyamoto, *An instability criterion for activator-inhibitor systems in a two-dimensional ball.* II, J. Differential Equations **239** (2007), 61-71.

[NS] T. Nagai and T. Senba, *Global existence and blow-up of radial solutions to a parabolic-elliptic system of chemotaxis*, Adv. Math. Sci. Appl. **8** (1998), 145-156.

[NSu] K. Nagasaki and T. Suzuki, *Radial and nonradial solutions for the nonlinear eigenvalue problem* $\Delta u + \lambda e^u = 0$ *as annuli in* $\mathbb{R}^2$, J. Differential Equations **87** (1990), 144-168.

[Na1] K. Nakashima, *Stable transition layers in a balanced bistable equation*, Differential Integral Equations **13** (2000), 1025-1038.

[Na2] K. Nakashima, *Multi-layered stationary solutions for a spatially inhomogeneous Allen-Cahn equation*, J. Differential Equations. **191** (2003), 234-276.

[Ne] Z. Nehari, *On a class of nonlinear second-order differential equations*, Trans. Amer. Math. Soc. **95** (1960), 101-123.

[N1] W.-M. Ni, *Recent progress in semilinear elliptic equations*, Publ. Res. Inst. Math. Sci. **679** (1989), 1-39.

[N2] W.-M. Ni, *Diffusion, cross-diffusion, and their spike-layer steady states*, Notices Amer. Math. Soc. **45** (1998), 9-18.

[N3] W.-M. Ni, *Qualitative properties of solutions to elliptic problems*, Stationary Partial Differential Equations **I**, 157-233, Hand. Differ. Equ., North–Holland, Amsterdam, 2004.

[NPY] W.-M. Ni, P. Polacik, and E. Yanagida, *Monotonicity of stable solutions in shadow systems*, Trans. Amer. Math. Soc. **353** (2001), 5057-5069.

[NT1] W.-M. Ni and I. Takagi, *On the Neumann problem for some semilinear elliptic equations and systems of activator-inhibitor type*, Trans. Amer. Math. Soc. **297** (1986), 351-368.

[NT2] W.-M. Ni and I. Takagi, *On the shape of least-energy solutions to a semilinear Neumann problem*, Comm. Pure Appl. Math. **44** (1991), 819-851.

[NT3] W.-M. Ni and I. Takagi, *Locating the peaks of least-energy solutions to a semilinear Neumann problem*, Duke Math. J. **70** (1993), 247-281.

[NT4] W.-M. Ni and I. Takagi, *Point condensation generated by a reaction-diffusion system in axially symmetric domains*, Japan J. Indust. Appl. Math. **12** (1995), 327-365.

[NTY1] W.-M. Ni, I. Takagi, and E. Yanagida, *Stability analysis of point-condensation solutions to a reaction-diffusion system*, Tohoku Math. J., submitted.

[NTY2] W.-M. Ni, I. Takagi, and E. Yanagida, *Stability of least-energy patterns in a shadow system of an activator-inhibitor model*, Japan J. Indust. Appl. Math. **18** (2001), 259-272.

[NTa] W.-M. Ni and M. Tang, *Turing patterns in the Lengyel-Epstein system for the CIMA reaction*, Trans. Amer. Math. Soc. **357** (2005), 3953-3969.

[NW] W.-M. Ni and X. Wang, *On the first positive Neumann eigenvalue*, Discrete Contin. Dyn. Syst. A **17** (2007), 1-19.

[NWe] W.-M. Ni and J. Wei, *On the location and profile of spike-layer solutions to singularly perturbed semilinear Dirichlet problems*, Comm. Pure Appl. Math. **48** (1995), 731-768.

[NY] W.-M. Ni and S. Yotsutani, *Semilinear elliptic equations of Matukuma-type and related topics*, Japan J. Indust. Appl. Math. **5** (1988), 1-32.

[Pn] X. Pan, *Surface superconductivity in applied magnetic fields above H_{c_2}*, Comm. Math. Phys. **228** (2002), 327-370.

[PW] L. Payne and H. Weinberger, *On optimal Poincaré inequality for convex domains*, Arch. Rational Mech. Anal. **5** (1960), 286-292.

[Pe] J. E. Pearson, *Complex patterns in a simple system*, Science **261** (1993), 189-192.

[Po] P. Polacik, *Parabolic equations: Asymptotic behavior and dynamics on invariant manifolds*, Handbook of Dynamical Systems **2**, 835-883, Hand. Differ. Equ., North–Holland, Amsterdam, 2002.

[PY] P. Polacik and E. Yanagida, *On bounded and unbounded global solutions of a supercritical semilinear heat equation*, Math. Ann. **327** (2003), 745-771.

[PS] G. Polya and G. Szegö, Isoperimetric Inequalities in Mathematical Physics, Ann. of Math. Stud. **27**, Princeton University Press, Princeton, NJ, 1951.

[PrW] M. H. Protter and H. F. Weinberger, Maximum Principles in Differential Equations, Prentice–Hall, Englewood Cliffs, NJ, 1967.

[R] X. Ren, *Least-energy solutions to a nonautonomous semilinear problem with small diffusion coefficient*, Electron. J. Differential Equations **no. 05** (1993), 1-21.

[Sa] D. G. Sattinger, *Monotone methods in nonlinear elliptic and parabolic boundary value problems*, Indiana Univ. Math. J. **21** (1972), 979-1000.

[S] O. Savin, *Regularity of flat level sets in phase transitions*, Ann. of Math. (2) **169** (2009), 41-78.

[Sc] D. G. Schaeffer, *An example of the plasma problem with infinitely many solutions*, preprint (1976).

[SZ] J. Serrin and H. Zou, *Symmetry of ground states of quasilinear elliptic equations*, Arch. Ration. Mech. Anal. **148** (1999), 265-290.

[SKT] N. Shigesada, K. Kawasaki, and E. Teramoto, *Spatial segregation of interacting species*, J. Theoret. Biol. **79** (1979), 83-99.

[Sh] S.-A. Shim, *Uniform boundedness and convergence of solutions to cross-diffusion systems*, J. Differential Equations **185** (2002), 281-305.

[Si] L. Simon, *Asymptotics for a class of non-linear evolution equations, with applications to geometric problems*, Ann. of Math. (2) **118** (1983), 525-571.

[Sm] J. Smoller, *Shock Waves and Reaction-Diffusion Equations*, Springer-Verlag, New York, Berlin, 1983.

[Sw] G. Sweers, *A sign-changing global minimizer on a convex domain*, Progress in Partial Differential Equations: Elliptic and Parabolic Problems, C. Bandle et al., eds., Pitman Res. Notes Math. Ser. **266**, Longman, Harlow (1991), 251-258.

[Ta] I. Takagi, *Point-condensation for a reaction-diffusion system*, J. Differential Equations **61** (1986), 208-249.

[Tl] S. D. Taliaferro, *Radial symmetry of large solutions of nonlinear elliptic equations*, Proc. Amer. Math. Soc. **124** (1996), 447-455.

[Tn] M. Taniguchi, *Multi-dimensional traveling fronts in bistable reaction-diffusion equations*, Discrete Contin. Dyn. Syst. A, to appear.

[Tr] A. Trembley, Memoires pour servir a l'histoire d'un genre de polypes d'eau douce a bras en forme de cornes, Leyden, 1744.

[Ty] W. C. Troy, *Symmetry properties in systems of semilinear elliptic equations*, J. Differential Equations **42** (1981), 400-413.

[T] A. M. Turing, *The chemical basis of morphogenesis*, Philos. Trans. Roy. Soc. London B **237** (1952), 37-72.

[Wm] P. Waltman, Competition Models in Population Biology, CBMS-NSF Regional Conf. Ser. in Appl. Math. **45**, SIAM, Philadelphia, 1983.

[Wa1] X. Wang, *On the Cauchy problem for reaction-diffusion equations*, Trans. Amer. Math. Soc. **337** (1993), 549-590.

[Wa2] X. Wang, private communication (2005).

[W1] J. Wei, *On the boundary spike layer solutions of singularly perturbed semilinear Neumann problem*, J. Differential Equations **134** (1997), 104-133.

[W2] J. Wei, *On the interior spike layer solutions to a singularly perturbed Neumann problem*, Tohoku Math. J. **50** (1998), 159-178.

[W3] J. Wei, *Pattern formations in two-dimensional Gray-Scott model: Existence of single-spot solutions and their stability*, Phys. D **148** (2001), 20-48.

[W4] J. Wei, *Conditions for two-peaked solutions of singularly perturbed elliptic equations*, Manuscripta Math. **96** (1998), 113-131.

[WW1] J. Wei and M. Winter, *Spikes for the two-dimensional Gierer-Meinhardt system: The weak coupling case*, J. Nonlinear Sci. **11** (2001), 415-458.

[WW2] J. Wei and M. Winter, *Spikes for the Gierer-Meinhardt system in two dimensions: The strong coupling case*, J. Differential Equations **178** (2002), 478-518.

[WW3] J. Wei and M. Winter, *Existence and stability of multiple-spot solutions for the Gray-Scott model in* $\mathbb{R}^2$, Phys. D **176** (2003), 147-180.

[WY] J. Wei and J. Yang, *Concentration on lines for a singularly perturbed Neumann problem in two-dimensional domains*, Indiana Univ. Math. J. **56** (2007), 3025-3074.

[Wi] N. Wiener, *Tauberian theorems*, Ann. of Math. (2) **33** (1932), 1-100.

[Wu] Y. Wu, *The instability of spiky steady states for a competing species model with cross diffusion*. J. Differential Equations **213** (2005), 289-340.

[Ya] E. Yanagida, *Mini-maximizers for reaction-diffusion systems with skew-gradient structure*, J. Differential Equations **179** (2002), 311-335.

[Y] S. Yotsutani, private communication (2009).

[Z] T. I. Zelenyak, *Stabilization of solution of boundary value problems for a second-order parabolic equation with one space variable* (in Russian), Diff. Urav. **4** (1968), 34-45.

Index